THE ANNALS OF
THE KING'S ROYAL RIFLE CORPS

JEFFERY, LORD AMHERST

(From an engraving after the portrait by Sir Joshua Reynolds, P.R.A.)

THE ANNALS

OF THE

KING'S ROYAL RIFLE CORPS

BY

LEWIS BUTLER

FORMERLY CAPTAIN IN THE REGIMENT

VOLUME I.

"THE ROYAL AMERICANS"

"To summon from the shadowy past
The forms that once have been."

WITH ILLUSTRATIONS AND MAPS

TO

HIS MAJESTY KING GEORGE V.

THE COLONEL-IN-CHIEF

THIS WORK IS WITH GRACIOUS PERMISSION

DEDICATED BY

THE AUTHOR

PREFACE

TO

THE FIRST VOLUME

To write the history of the King's Royal Rifle Corps demands infinitely greater literary ability than that of the present writer, and infinitely more time than he can spare. Ten years of unremitting labour might hardly suffice, for although plenty of material may be in existence it must be searched for not only in such places as the British Museum and the Record Office, but across the seas in distant countries; and even there, not only in official libraries, but among family papers and private documents, the value of which is probably quite unknown to the owners.

Nor can it be denied that the Regiment itself has been much to blame for the laxity with which its historical records have been kept. The oldest regimental works of reference extant are the History of Services of the 1st Battalion, and the Records of the 2nd Battalion, of which, however, the former appears to have been taken in hand about 1823, and the latter in 1831.[1] Although this book of the 1st Battalion makes some attempt at giving an outline of the regimental history from 1755, the result is

[1] Writing in that year, Colonel Ellis refers to the existence of a Record Book begun in 1808. It has no doubt long since disappeared.

not very satisfying, and consists of little but extracts from the 'Memoir of Major Patrick Murray.'

Dealing with matters of more recent date these records have a certain value, but it constantly happens that the dry bones of facts—sometimes of purely ephemeral interest, sometimes of importance—are given without any explanation of the circumstances under which those facts took place, which alone would make them of value to the historian. For instance, under the date October 13, 1881, the History of Services of the 1st Battalion states that 'the Bn. under command of Lieut.-Colonel Dundas proceeded by route march to Sallins in aid of the Civil power, returning to camp same day.' From this bald statement it would hardly be gathered that so serious was the condition of Ireland that the Government decided to arrest the well-known leader, Charles Stewart Parnell; and was so apprehensive of armed interference, even at a wayside railway station, that it sent two brigades, one of cavalry and one of infantry, from the Curragh to enforce its orders.[1] In this respect the records of the 2nd Battalion are on the whole better than those of the 1st, but both leave something to be desired.

In the Regiment the credit of keeping alive the lamp of historical narrative has, until lately, been due almost exclusively to three officers: the late Major-General Gibbes Rigaud, Major-General Astley Terry, and Colonel N. W. Wallace; and to each of these three the Regiment is greatly indebted. In the year 1879 the last-named published 'A Regimental Chronicle and List of Officers.'[2] This work was, at the last, perhaps somewhat hurriedly

[1] As it turned out the display of force was unnecessary, for Parnell was arrested in Dublin.

[2] Colonel Wallace's book is now out of print; it would be in the historical interests of the Regiment to publish a new edition, corrected, indexed, and revised according to the requirements of the present day.

brought out, and suffered from lack of revision; nevertheless, it has been of great help in the compilation of the present volume.

The compilation of the present work was due to the initiative of the late Sir Redvers Buller, who furnished the greater part of the necessary funds, and requested the Hon. John Fortescue, the ablest military historian of the day, to undertake the task. Mr. Fortescue complied, but after bringing a draft of the narrative down to the year 1802, found himself compelled by the pressure of other work to relinquish it. He then most generously presented the MS. to the Regiment, giving the present writer free permission to make what use of it (with or without addition, omission, or alteration) he pleased. Of this permission full advantage has been taken; but the document forms the basis of much of the present volume. It is hardly possible for the Regiment sufficiently to acknowledge its debt to Mr. Fortescue; without his work, and without that of Colonel Wallace, it would have been impossible for the present writer to attempt the task.

During the period of which the first volume treats, only one private memoir has been available, namely that of Major Murray (given in Appendix II). This memoir was, as we are told by the History of Services of the 1st Battalion, handed by the author to the Officer Commanding the Battalion, probably Colonel Andrews, at Montreal in 1822. The original is at present in the hands of Mr. H. J. Bunbury, a descendant of Colonel Thomas Bunbury, who succeeded Andrews in the command of the 1st Battalion, but it seems probable that the donor intended it to be the property of the Regiment. The memoir for the most part describes the experiences of the author in the American War of Independence; it shows lack of revision, and terminates abruptly, but is of considerable interest, not only in throwing light on the Regiment of that day, but as being the work of an officer who must have been familiar with many original members of the 'Royal Americans.'

For the period from 1756 to 1764 the reader is referred to Parkman's delightful volumes entitled 'Montcalm and Wolfe' and 'The Conspiracy of Pontiac'; and something of historical interest may be gleaned from the works of J. Fenimore Cooper, notably 'The Last of the Mohicans.' Cooper married a Miss Delancey, believed to be the daughter of an officer of the 60th who was present at some of the scenes described by his son-in-law.

It is, perhaps, unnecessary for me to refer to Sir Edward Hutton's excellent monograph on Henry Bouquet, which has no doubt been read by every member of the Regiment.

My acknowledgments are due to many others too numerous to mention by name, and not least to Mr. Ernest Christie Brown of New York, who astounded me by his intimate acquaintance with the careers of the officers of the Royal American Regiment in 1756, and presented me with Parkman's works.

This volume attempts to describe the careers of the 1st, 2nd, 3rd, 4th, and 6th Battalions up to the year 1815, leaving those of the 5th, 7th and 8th to a future time. The lists and tables given in the Appendix are taken from the sources of information to which the writer has had access. He will not, however, answer for their absolute completeness or accuracy.

The events described precede the date at which the Regiment was officially recognised as a Rifle Corps, but if the writer were to be asked his views on the attributes of the ideal rifleman, he would reply that the latter should unite in his own person the qualities of Bouquet,

Augustin Prevost, de Rottenberg, Hawley, and Buller; and this combination, so far as he knows, is to be found only in one man, who, as it happened, did, for a few months, serve in our Regiment—Sir John Moore.

With these words and with many apologies for the innumerable shortcomings of the work, the writer ends his preface. No one realises better than himself how little he knows of the history of the Royal American Regiment, and how greatly he has failed to do justice to it; yet if he has done anything to arouse interest in its early days and to recall names and episodes

'softly fading in oblivion'

he will be satisfied; in the hope that some abler man, in the not distant future, will write the book which may fairly claim to be called the History of the Regiment.

Only one word more. While these pages were in the press, the greatest soldier and the most far-seeing reformer of our age passed away. To the world at large Lord Wolseley had been for several years lost; but it may interest some to hear that to the end he recalled with pleasure episodes of the days in Canada when he was on intimate terms with our Regiment—particularly with the 1st Battalion which served under his command in the Red River Expedition—and that when, as occasionally happened, his opinion was asked in regard to some passage in this volume he never failed to give it in the kindest and also clearest possible manner.

LEWIS BUTLER.

WATTON HOUSE, HERTFORD:
May 12, 1913.

CONTENTS

THE INDIAN REVOLT (1763–1764)

CHAPTER VI

CHAPTER VII

CHAPTER VIII

THE AMERICAN WAR OF INDEPENDENCE (1775–1783)

CHAPTER IX

THE GREAT WAR (1793–1815)

CHAPTER X

CHAPTER XI

CHAPTER XII

CHAPTER XIII

APPENDICES

ILLUSTRATIONS

PORTRAITS

MAPS

REGIMENTAL SUCCESSION OF COLONELS-IN-CHIEF, COLONELS-COMMANDANT, AND LIEUTENANT-COLONELS, 1755–1798

COLONELS-IN-CHIEF

John, Earl of Loudoun	Dec. 25, 1755.
James Abercrombie	Dec. 27, 1757.
Jeffery Amherst	Sept. 30, 1758.
Thomas Gage	Sept. 27, 1768.
Sir Jeffery Amherst, K.B. (2nd time)	Nov. 7, 1768.
H.R.H. Frederick, Duke of York, K.G.	Aug. 23, 1797.

COLONELS-COMMANDANT

Batt.		
1.	John Stanwix	Jan. 1, 1756.
2.	Joseph Dusseaux	Jan. 2, 1756.
3.	Charles Jeffereys	Jan. 3, 1756.
4.	Jacques Prevost	Jan. 4, 1756.
3.	George, Viscount Howe	Feb. 25, 1757.
3.	Charles Lawrence	Sept. 28, 1757.
2.	Hon. Robert Monckton	Dec. 20, 1757.
2.	Hon. James Murray	Oct. 24, 1759.
3.	William Haviland	Dec. 9, 1760.
4.	Marcus Smith	Nov. 11, 1761.
2.	Bigoe Armstrong	Dec. 16, 1767.
2.	Frederick Haldimand	Oct. 20, 1772.
3.	William Taylor	Sept. 17, 1775.
4.	Augustin Prevost	Sept. 18, 1775.
2.	James Robertson	Jan. 11, 1776.
3.	John Dalling	Jan. 16, 1776.

BATT.		
2.	Gabriel Christie	May 14, 1778.
3.	William Rowley	Oct. 3, 1787.
4.	Hon. William Gordon	Oct. 3, 1787.
4.	James Rooke	Oct. 20, 1788.
1.	Alured Clarke	July 8, 1791.
1.	Thomas Carleton	August 6, 1794.
4.	Peter Hunter	August 2, 1796.

SUCCESSION OF COLONELS-COMMANDANT

BY BATTALIONS

1st Battalion

John Stanwix	Jan. 1, 1756.
Jacques Prevost	Oct. 28, 1761.
Frederick Haldimand	Jan. 11, 1776.
Alured Clarke	July 8, 1791.
Thomas Carleton	Aug. 6, 1794.

2nd Battalion

Joseph Dusseaux	Jan. 2, 1756.
Hon. R. Monckton	Dec. 20, 1757.
Hon. J. Murray	Oct. 24, 1759.
Bigoe Armstrong	Dec.16, 1767.
Frederick Haldimand	Oct. 20, 1772.
James Robertson	Jan. 11, 1776.
Gabriel Christie	May 14, 1778.
Augustin Prevost	—— 1783.
Gabriel Christie	May 10, 1786.

3rd Battalion

Charles Jeffereys	Jan. 3, 1756.
George, Viscount Howe	Feb. 28, 1757.
Charles Laurence	Sept. 28, 1757.
William Haviland	Dec. 9, 1760.
William Taylor	Sept. 17, 1775.
John Dalling	Jan. 16, 1776.
William Rowley	Oct. 3, 1787.

4th Battalion

Jacques Prevost	Jan. 4, 1756.
Marcus Smith	Nov. 11, 1761.
Augustin Prevost	Sept. 18, 1775.
Hon. Wm. Gordon	Oct. 3, 1787.
James Rooke	Oct. 20, 1788.
Peter Hunter	August 2, 1796.

Note.—It will be observed that Haldimand, A. and J. Prevost were Colonels-Commandant—at different times—of more than one battalion. When the 4th Battalion was reduced in 1783, A. Prevost was given Christie's battalion; but on Prevost's death in 1786, it was handed back to Christie.

SUCCESSION OF LIEUTENANT-COLONELS

1st Battalion

Henri Bouquet	Jan. 3, 1756.
Augustin Prevost	Dec. 13, 1765.
Gabriel Christie	Sept. 18, 1775.
Stephen Kemble	May 14, 1778.
George Etherington	—
James A. Harris	Jan. 16, 1788.

2nd Battalion

Frederick Haldimand	Jan. 4, 1756.
Sir John St. Clair	—
Frederick Haldimand	—
Gabriel Christie	August 14, 1773.
George Etherington	Sept. 19, 1775.
Stephen Kemble	—
Hon. Vere Paulet	May 13, 1793.

3rd Battalion

Russell Chapman	Jan. 5, 1756.
John Young	April 26, 1756.
Augustin Prevost	March 20, 1761.

William Stiell . . . . Sept. 21, 1775.
Archibald McArthur . . . . . . April 24, 1781.

Archibald McArthur . . . . . . Sept. 24, 1787.
William G. Strutt . . . . Nov. 18, 1790.
Gerrit Fisher . . . . April 25, 1792.
John Ritchie . . . . Feb. 17, 1794.
Hon. John Elphinstone . . . . . . July 20, 1794.

4th Battalion

Sir John St. Clair . . . . Jan. 6, 1756.
Frederick Haldimand . . . . . . —

Lewis Val. Fuser . . . . Sept. 20, 1775.
Beamsley Glazier . . . . May 6, 1780.

Peter Hunter . . . . Sept. 24, 1787.
George Prevost . . . . August 6, 1794.

SYNOPSIS OF CONTEMPORARY EVENTS

1755-1800

1755, July .	. Defeat of General Braddock on the Monongahela.
1756 . .	. Beginning of the Seven Years' War. England and Prussia *v.* France, Austria and Russia.
1757 . .	. Supremacy of the elder Pitt.
May 6	. Frederick the Great, King of Prussia, defeats the Austrians at Prague.
June 18	. Frederick defeated at Kolin.
June	. Colonel Clive wins the battle of Plassey and establishes British power in Bengal.
Sept. .	. Duke of Cumberland defeated at Hastenbeck. Concludes the Convention of Kloster Seven and is consequently superseded by Sir John Ligonier.
Nov. 5	. Frederick defeats the French at Rossbach, and
Dec. 5	. The Austrians at Lenthen.
1758, April	. British troops sent to Germany under command of Prince Ferdinand of Brunswick.
Aug. 25	. Frederick defeats the Russians at Zorndorf, but
Oct. 13	. Is defeated by the Austrians at Hochkirch.
1759, Aug. 1	. Ferdinand defeats the French at Minden.
Aug. 12	. Frederick defeated by the Russians at Kunersdorf.
Nov. 10	. Admiral Hawke defeats the French fleet at Quiberon Bay.
1760, Oct. .	. Death of George II.
1761, Jan. 17	. Capture of Pondicherry from the French.
Oct. 5	. Fall of Pitt.
1762, Jan. 2	. England declares War against Spain.
1763, Feb. 10	. Peace of Paris between France, Spain and England.
1776, July 4	. Declaration of Independence by the United States of America.

1778, Oct. . .	Second capture of Pondicherry.
1779–1783 . .	Successful defence of Gibraltar by General Elliot.
1782, April 12 .	Admiral Rodney defeats the French fleet near Guadeloupe.
1783, Sept. 3 .	Peace of Versailles.
Dec. 23 .	William Pitt the younger becomes Prime Minister.
1789, May 5 .	Meeting of the States General at Versailles.
July 14 .	Capture of the Bastille. Beginning of the French Revolution.
1793, Jan. 21 .	Execution of Louis XVI.
Feb. 1 .	France declares War on England.
	Campaign in Flanders.
Dec. 19 .	The British and Spaniards driven from Toulon.
1794, June 1 .	Naval victory of Lord Howe.
1796 . . .	General Bonaparte's victorious campaign in Italy.
1797 . . .	Mutiny of the British fleet.
Feb. 14 .	Admiral Jervis defeats the French and Spanish fleets off Cape St. Vincent.
Oct. 11 .	Admiral Duncan defeats the Dutch at Camperdown.
1798–1799 . .	Bonaparte's campaign in Egypt and Syria. Rebellion in Ireland. General Humbert lands in Ireland with a French force but is compelled to surrender.
1799, Sept.–Oct. .	Anglo-Russian campaign in the Helder. British evacuate Holland. General Bonaparte elected First Consul in France.
1800, June 14 .	Bonaparte defeats the Austrians at Marengo.
1801, March .	Sir R. Abercromby lands with a British Army in Egypt and beats the French at the battle of Aboukir. Evacuation of Egypt by the French.
1802, March 27 .	Peace of Amiens.

NORTH AMERICA
English Miles
0 50 100 200 300 400 500
Spanish circa 1700
Dutch (British 1664)
Swedish (Dutch 1655, British 1664)
French before the Peace of Utrecht 1713
British before the Peace of Utrecht 1713
Territory in dispute between French & British before the Peace of Utrecht 1713
(British 1713)
Newfoundland (British 1713)
Gulf of St. Lawrence
L. Superior
L. Michigan
Huron
L. Ontario
L. Erie
NEW YORK
PENNSYLVANIA
MASSACHUSETTS
RHODE ISLAND
NEW JERSEY
DELAWARE
VIRGINIA
NORTH CAROLINA
SOUTH CAROLINA
GEORGIA (British 1733)
WEST FLORIDA (British 1763-1781)
EAST FLORIDA (British 1763-1783)
LOUISIANA
ATLANTIC OCEAN
GULF OF MEXICO
Quebec
Montreal
Three Rivers
Halifax
Louisbourg
Boston
Plymouth
Long Island
New Amsterdam (New York)
Baltimore
Washington
Yorktown
Charleston
Savannah
St. Augustine
Pensacola
Mobile
New Orleans
Mouths of the Mississippi
R. Mississippi
R. Missouri
Ohio R.
Cumberland R.
R. Tennessee
Ft. Duquesne (Ft. Pitt)
Ft. Detroit
Ft. Niagara
Ft. Frontenac
Toronto
Sault Ste. Marie
Michillimackinac
Vincennes
Ft. Chartres
Nova Scotia (British 1713)
Port Royal (Annapolis)
B. of Fundy
Cape Breton I. (Ile Royale)
Pr. Edward I. (Ile de St. Jean)
C. Cod
Emery Walker sc.
The West Indies and Central America
English Miles
0 100 200 300 400 500 600
GULF OF MEXICO
Bahama Islands
Havana
Cuba
Jamaica
Port Royal
Haiti
St. Domingo
Hispaniola
Porto Rico
St. Thomas
Santa Cruz
St. Martin
St. Kitts
Nevis
Montserrat
Antigua
Guadeloupe
The Saints
Dominica
Martinique
St. Lucia
St. Vincent
Barbados
Grenada
Tobago
Trinidad
CARIBBEAN SEA
ATLANTIC OCEAN
PACIFIC OCEAN
Yucatan
Honduras
Nicaragua
Mosquito Coast
C. Gracias á Dios
L. Nicaragua
Fort St. Juan
R. St. Juan
Porto Bello
Panama
Darien
Cartagena
R. Orinoco

THE KING'S ROYAL RIFLE CORPS

INTRODUCTION

GENERAL BRADDOCK—CONSEQUENCES OF HIS DEFEAT—THE AMERICAN COLONIES—THE ROYAL AMERICAN REGIMENT RAISED

'WE shall learn better how to do it the next time.' Such was the last exclamation of General Braddock, who had been borne away from the fatal field by the Monongahela River, only to die a day or two later by the roadside. A good and capable officer, even though of the 'sealed pattern' order, he had by no means underrated the difficulties of his enterprise. He had triumphed over many. He had complied with all the instructions laid down in the Field Exercises of the day, but he had not foreseen that the regulations drawn up for manœuvring in one hemisphere might be quite unsuitable in another: that the tactics which had made troops victorious over the best armies of Europe might of themselves lead to annihilation by a Franco-Indian force in America. Despite his own conspicuous bravery, the General had seen his army disorganised under the fire of an invisible foe and his men—fine, trained soldiers—reduced to the state of a helpless mob.[1] But now as life ebbed rapidly away, he thought no longer of the terrible scene of defeat, of his ruined reputation, or of the sneers and aspersions that would be made on his memory. A gallant confidence once more asserted itself. He himself had failed; but another, profiting by his defeat, would grasp the situation, and by dint of new and improved methods would avenge his death and carry out the far-reaching plans of his strategy. 'We shall learn better how to do it the next time.'

Whether these words were reported to H.R.H. the Duke of Cumberland, at this period (1755) Commander-in-Chief, is unknown; but certain it is that he at once grasped the full significance of Braddock's defeat and of other disasters which had preceded it. Regulation had been strictly adhered to—even when, during the heat of the action, the British soldiers, in imitation of the provincial levies, had tried to seek cover, Braddock had forced them back into the open: regulation had proved a broken reed. It was obvious that regulation must be set aside and a corps organised which could learn to compete successfully with the French and their Indian allies, and defeat them by superior proficiency in their own methods.

According to Major Patrick Murray,[2] an officer who served in the 60th from 1770 to 1793, M. von Harbot, a Swiss belonging to the Canton of Berne, heard of Cumber-

[1] 'British officers,' writes Bradley (*The Fight with France*, on pp. 100–1), 'have declared that no pen could describe the scene. One actor in it wrote that the dreadful clangour of the Indians' war-whoop would ring in his ears till his dying day. . . . Everything was abandoned to the enemy. . . . It became a race for life. . . . Regulars and provincials splashed in panic and dire confusion through the ford they had crossed in such pomp but three hours before. Arms and accoutrements were flung away in the terror with which men fled from these ghastly shambles. . . . Scores of helpless wounded were left victims to the tomahawk and the scalping knife.'

[2] *Vide* Appendix II.

land's difficulty and made a proposal to his countryman, M. Jacques Prevost. This was to raise a regiment of four battalions, largely composed of provincials, in America under a British colonel-in-chief, but with a fair sprinkling of foreign officers. Inasmuch, however, as Prevost, a soldier of fortune, was serving in Holland, where the sister to the Duke of Cumberland was the Princess of Orange, it seems more than probable that Prevost's sympathies were enlisted through the medium of the Princess and the British Minister at The Hague. In any case the proposal was quickly adopted and the task of forming the cadre assigned to Prevost, who set himself to work with great vigour and spared no pains to collect for the new regiment the best officers available. Henri Bouquet and Frederick Haldimand, Swiss officers of distinction serving in the bodyguard of the Prince of Orange, were appointed to the lieutenant-colonelcies of the 1st and 2nd Battalions. Prevost's elder brother Augustin was made a major in the regiment; and his younger brother Jean Marc, a captain. These last had hitherto been serving in Holland as their brother's captain-lieutenant and lieutenant respectively. A more brilliant quartette has rarely been found in any regiment: and of the first name it may be said that although in the 60th we have had other commanding officers of the first rank, there has been none to surpass, none perhaps even to equal, Henri Bouquet.

To form a body of regular troops capable of contending with the Red Indian in his native forest by combining the qualities of the scout with the discipline of the trained soldier, was the object in raising the Royal American Regiment; and from the first day of its existence until the present time that combination has been its characteristic. The idea of dressing such soldiers in anything but the universal red uniform, or of arming them with anything but the smooth-bore weapon of the period, had not in 1755 been formed. But, from the very outset, the regimental system lay in adapting the rôle of the 'hunter' to the military needs of the day by cultivating the woodcraft and physical hardihood of the backwoodsman.[1] The result was the creation of that special type of fighting man known to the Frenchman as the 'Chasseur'; to the German as the 'Yäger'; to ourselves as the Scout or Rifleman. And the art of the scout, be it remembered, was learnt, not theoretically, within the safe confines of an English camp of exercise, but in practice, among the hill and valley of the primeval forest destitute almost of pathway, and whose every tree or undulation might be the ambush of the Red-skin, himself as fine a model of a scout as the world has seen.

In forest warfare the full dress uniform was quickly discarded in favour of more suitable and less conspicuous tones: with the rifle[2] the men of the Royal American Regiment had, as it turned out, for the most part been long familiar, for it was the usual weapon of the backwoodsman; nevertheless at a later date it was not the change in the colour of its uniform nor the adoption of a superior weapon, but the strenuous training as scouts and skirmishers in which from the outset of their career the men of the 60th had been exercised, and the process of ordinary evolution, that entitled the Regiment to become in name, as for nearly seventy years it had essentially been in fact—a Rifle Corps.

The regiment raised under the name of the Royal Americans—afterwards known as the 60th Rifles, and at the present day termed the King's Royal Rifle Corps—

[1] *Vide* Bouquet's plan for raising battalions of 'Hunters,' p. 160 and Appendix IV.

[2] The word 'rifle' shows its German origin, being derived from *riefeln*—'to groove.'

has had its good and its bad times, its failures as well as its successes. It has no wish to exaggerate success or to minimise failure; but whatever the result of an episode in its history, whether good or bad, the object of the Regiment has always been to vindicate the *raison d'être* of its existence and to make good the dying General's words in the determination to 'learn better how to do it the next time.'

But before going further a word of explanation must be given as to the position of England in America.

After many ineffectual attempts the first permanent British settlement had been made in 1607 on the James River in Virginia. Thirteen years later a British colony was founded at Plymouth in Massachusetts. Meanwhile the Dutch had established settlements on the east coast and in 1614 had founded the city of New Amsterdam on the Hudson River. In 1664 the British being at war with the Dutch seized their settlements, and changed the name of New Amsterdam to New York in honour of the King's brother, the Duke of York, afterwards James II.

A vast and most important territory was thus added to the land already occupied to the north and south by the British settlers; and the Duke of York, to whom the King granted the new province, rightly decided to provide it with a garrison, at first of two, and later of four companies. It is only fair to say that the American colonies for the most part made at first no excessive demand upon the mother country for military aid. The settlers in the various provinces had carried with them the English system of Militia, which long sufficed for all their needs. But the situation became gradually altered. A neighbour and rival had appeared in the north. So far back as 1534 Jacques Cartier,[1] the French navigator, had sailed up the St. Lawrence and gained a footing on the shore. In 1608 his countryman,[2] Samuel Champlain, founded Quebec, which though captured by the British in 1629 was restored three years later; and thenceforward the French power in Canada, or New France as it was usually called, grew so rapidly as to become a menace to the English colonies.

'The occupation by France of the valley of the St. Lawrence with its dependencies, such as Newfoundland, Cape Breton and Acadia,' observes Mr. Fletcher,[3] 'was on the whole more intelligently conceived than the settlement of British America. The French colony was administered as a whole, and was governed and well equipped from a military point of view; great statesmen

[1] Jacques Cartier, born at St. Malo, 1491: heard in boyhood of the newly discovered continent and in youth sailed with the fishing fleet year by year to Newfoundland. In 1534, at the instance of Chabot, High Admiral of France Cartier, now a master-pilot, was selected by the King, Francis I, to command an expedition projected to find a passage to the continent of Asia by the American coast. Towards the end of July, Cartier discovered the mouth of the St. Lawrence, and hoped he had reached the desired channel; but stress of weather compelled him to desist and return home. Next year he sailed past the modern Quebec amid the welcome of the Indian tribes as far as their settlement of Hochelaga, where he landed and named the neighbouring eminence the Royal Mountain—Montreal. It was now obvious that the water on which he was sailing was merely a river, and could not possibly take him to Asia, and he returned to France in disappointment and with no conception of the fact that he had indeed discovered the first stage of the way thither.

[2] S. Champlain, born in 1574: in 1603 ascended the St. Lawrence River up to the falls of St. Louis, the farthest point reached by Cartier in 1535. After founding Quebec, he formed a confederacy of the Huron and Algonquin Indians against the Iroquois, who were thenceforward forced to seek alliance with the British. In 1615 Champlain penetrated to Lake Ontario and spent a winter among the Hurons. In 1629 he was compelled to surrender Quebec to a British naval force commanded by a French refugee named Sir D. Kertk, but by the treaty of St. Germains Quebec was restored to France in 1632. Two years later he was appointed Governor of New France, but died shortly after arrival. The well-known lake in the State of Vermont, U.S.A., perpetuates his name.

[3] *Introductory History of England*, vol. iii. p. 189.

such as Richelieu,[1] Colbert[2] and Seignelay devoted attention to it; great pioneers like Frontenac,[3] La Salle,[4] and La Gallissonnière,[5] were its governors; and these perhaps foresaw, as no English statesmen did, that America would one day be a battle-ground between the two nations. The despotic government was powerful enough to override mercantile and provincial jealousies; and, while British colonists were often on the edge of rebelling, the population of Canada was always loyal to the French Crown. France also enforced better treatment of the Red Men, as she did of her slaves in the West Indies, whereas no British Governor dared to interfere with the sacred right of a free-born Briton to flog his own property. . . . In an age when Protestant missions to the heathen were all but unknown, the French Jesuits showed the same untiring zeal and devotion in the New World as in China. The gayer nature of the French people led them to drink, dance and intermarry with the natives—even to dress,

[1] Richelieu, Armand J. de P., Cardinal, Duc de, 1585–1642, first Minister of France under Louis XIII.

[2] Colbert, J. B., 1619–1683. A distinguished Minister of Finance and the Navy under Louis XIV.

[3] Frontenac, Louis de Bouade, born 1620; entered the army in 1635; selected by the great Turenne in 1669 to command the Venetian troops in the defence of Crete. Being successful he was appointed Governor of New France in 1672, in order, said the French wits, to give him an opportunity of getting away from his wife. Showed at Quebec great powers of administration and political ability, but was recalled to France in 1682. In 1689 he was sent back to Quebec in order to resist Indian encroachment instigated by the English. Hardly had he landed when he was attacked by a British force whom he successfully repelled. He died in 1698.

[4] Robert La Salle, born at Rouen about the year 1635. At the age of thirty-two went to Canada. In 1681 embarked on Lake Michigan and followed the course of the Mississippi and Illinois through unexplored forests until in April he reached the Gulf of Mexico, and called its shores Louisiana in honour of his sovereign. A few years later, during another tour of exploration, he was murdered by two of his own men. A monument to La Salle is to be seen in Rouen Cathedral.

[5] Roland, Comte de la Gallissonnière, a naval officer; acted as Governor of New France from 1747 to 1749. Returning to Europe in 1756, it was he who encountered the fleet of the unfortunate Admiral Byng, and forced it to retire. Byng was shot by sentence of court-martial; La Gallissonnière died the same year. His greatest talents were displayed in the fields of administration and science.

as Frontenac once did, in their war-paint and feathers—but the dour Briton thought it shame to do such things, which he classed with the ritual of Popery, so appealing to the Indians, as "monkey tricks." The result was that the French won the sympathy, and often the terrible help of the "forest-children," as our people never did, although, when all North America was ours the Red Men came to realise that only the British Government could protect them against the British colonist.'

Though in number far exceeding the French in North America, the British settlers, from internal dissension, could never combine to crush their enemies. The French, united under a single military governor, with a regular force of from one thousand to fifteen hundred men, were more than a match for any one single British colony. Narrow-minded, cantankerous, and full of local jealousies, the various provinces were eternally quarrelling among themselves, and were far more inclined to take up arms against each other from time to time than to work heartily for the good of all.[1] New York was the frontier-state towards Canada; and, bearing in mind that she possessed a small garrison paid for by the British Government, the remaining provinces were well content to take shelter behind her. As a matter of fact New York counted for protection chiefly upon the Indian tribes known as the Five Nations or Iroquois, whose friendship as an exceptional case the British had long cultivated, and with whom a solemn treaty of alliance was annually renewed. Upon these unhappy natives fell for many years the brunt of every encounter between British and French in North America, and they have paid for their fidelity by their extinction.

[1] Massachusetts formed an exception. Her citizen soldiery was generally ready to respond to the defence, not only of that state, but of the sister colonies.

When in 1689, upon the acceptance of the British throne by William III, Louis XIV of France declared war against England, the Government of Canada chanced to be in the hands of an extremely able and energetic man, Count Frontenac. The Revolution in England had extended itself to America; and in certain of the colonies, notably in New York, there was great disorder, amounting for a time almost to anarchy. These troubles intensified the normal inefficiency of the Colonial Governments, while at the same time political and religious animosity embittered the jealousies which divided province from province and party from party within the provinces themselves. The military precautions taken by the Governors of the late King, James II, were undone by the Revolutionists, simply because they were the work of James; and Frontenac seized the opportunity to carry fire and sword, not only into New York, but into New England. Massachusetts, always jealous of British power, made a wild attempt to capture Quebec unaided, but was foiled with disgrace as well as defeat; and Frontenac was not slow to punish her for her temerity. The brunt of his attack fell upon New York, but very soon Massachusetts which had hitherto never asked help of England, was also driven to implore the aid of the mother country in driving the French, 'the fountain of all our miseries,' from Canada. Had Frontenac received from King Louis the reinforcement which he asked, he would have captured New York, and probably the whole of the British possessions in North America. As events turned out, he was unable to penetrate far into British territory; but he wrought enormous mischief and destruction, and the power of the Five Nations received a shock from which it never fully recovered.

So far King William had done his best to help the

colonies by ordering every province to furnish a quota of men or a contribution in money to New York; but his commands were unheeded, and New York was left to shift for herself. The King thereupon answered the appeal of Massachusetts by sending out an expedition, which, by general mismanagement both in England and on the Atlantic, accomplished nothing; and at the Peace of Ryswick in 1697, the honours of the contest lay decidedly with Frontenac.

Simultaneously the French began to threaten the British settlements with a new danger. Unlike the British colonists, whose industry was almost exclusively agricultural, they sought rather to make a livelihood by trading with the Indians for fur; which preference, added to a natural spirit of enterprise, led them deep into the heart of the country. It was indeed, to Frenchmen that the first exploration of the gigantic continent of North America was due. In 1613 Samuel Champlain led the way to Lake Huron, and was presently followed by Jesuit missionaries, men of indefatigable energy and devoted courage, who spread rapidly to Lakes Superior and Michigan, taking possession of vast tracts of land in the name of King Louis XIV. Before Charles II had been long on the English throne, a French officer had conceived the plan of getting into the hinterland of the British settlements, and confining them to a narrow strip along the seaboard. A young adventurer, Robert La Salle, took in 1670 the first step towards consummating this policy by discovering the strait of Detroit, which leads from Lake Huron to Lake Erie, and pushing on to a branch of the Ohio. A Jesuit, in 1673, struck the Mississippi by way of the Wisconsin, and followed it down to the junction of the Arkansas. Finally, during the year 1681, in a marvellous expedition, La Salle penetrated from the present site of Chicago to the northern branch of the Illinois, paddled down the Mississippi, and pursued the course of the mighty river to its mouth in the Gulf of Mexico—a voyage of 5000 miles.

With the knowledge thus gained the French lost no time in erecting forts at Niagara, at Machinaw on the strait between Lakes Erie and Huron, and at other strategic points, in order to secure their new possessions. Moreover, then as now, they made with business-like forethought and careful industry maps of the country which they coveted. This was a sure method of getting the better of the British in any dispute over boundaries; for the British, having no maps of their own, could not impugn the accuracy of their neighbours' surveys. Hand in hand with this insidious undermining of the British position in North America, open hostilities were practically incessant. In 1703 the War of the Spanish Succession brought France and England again into conflict; and while Marlborough was winning his great victories in Europe, there was no lack of fighting in the New World. Once again, in 1711, a great effort was made to drive the French from Canada; but the expedition sent for the purpose from England was dogged by misfortune and incapacity from beginning to end, and accomplished literally nothing. At last, in 1714, came the Peace of Utrecht, which brought rest to Europe but not to America. Under its provisions the province of Acadia was ceded by France to England, but the boundaries of Acadia [1] had never been settled; and the French had no idea of allowing so good a cause of quarrel to perish. They kept the British frontier in constant alarm by stirring up

[1] Acadia is generally treated as synonymous with Nova Scotia, but it included the whole of that province and more. To the date of its disappearance it was one of the vaguest and most elastic of geographical terms.

the Indians against the settlers; and conceived the ingenious idea of restricting British territory by the establishment of a line of mission stations, which, owing to the influence of the Jesuits with the Indians, was practically a chain of military posts. This was too much. In 1724 the British lost patience, and swept away these stations by force; but, in the meanwhile to the north and east of our colonies the French had been equally busy in building still more important posts which were not so easily to be rooted out: and in order to follow the story of the ensuing operations the positions of these posts should be carefully noted and borne in mind.

The most memorable of them was the fort of Louisbourg, built upon a haven in the island of Cape Breton, in order to serve as a base for future aggression against New England and Nova Scotia. Louisbourg was a costly folly. The climate was such that no masonry could be kept in repair; and moreover, since the island provided nothing, the fortress was dependent upon the sea for its supplies, and was therefore useless without maritime supremacy. A new fort at Niagara was a more serious matter, as was likewise another stronghold erected at Chambly to cover Montreal from any attack by way of Lake Champlain. But the most aggressive fortress of all was a massive structure of masonry at Crown Point, the southern entrance of Lake Champlain, for it was actually built upon the territory claimed by New York. The province of New York, however, was so much engrossed by a squabble with her sister province of New Jersey, that she could spare neither time, money, nor men to oust this dangerous neighbour; and Crown Point was allowed to grow up without further molestation than a salvo of wordy protests. The only effort to counter these moves of the French was made by Governor Burnet of New York, who built the fort at Oswego on Lake Ontario as a rival to Niagara. But this he did at his own expense; and the debt due to him by the province on this account has never been repaid to this day.

So matters drifted on during the twenty years of nominal peace which distinguished Walpole's administration. Then at last in 1739, England became at variance with Spain, and in 1741 a great expedition was undertaken against the Spanish colonies in South America. After so long a period of peace England, of course, possessed few troops fit for war, and it was necessary to raise hasty levies both at home and in the American colonies. This last, it must be observed, was no new thing. Oliver Cromwell, for instance, had drawn troops from New England for his attack upon Jamaica in 1654; and it was with little difficulty that in 1740 Governor Spotswood of Virginia raised a regiment of four battalions, which, curiously enough, ranked as the Sixtieth of the Line. The expedition, however, ended disastrously with the practical annihilation of the British force, and Spotswood's battalions were speedily dissolved.

Within a year England was again at war with France over the question of the Spanish succession; and, while her troops in Europe were busy with such actions as Dettingen and Fontenoy, the eternal struggle on the Canadian frontier was renewed. The French, having first information of the outbreak of hostilities, were able to take the initiative in attack; but on this occasion the colonists, irritated beyond endurance, did not wait for a force from England to help them in taking their revenge. With the aid of a British fleet only, they boldly invested Louisbourg in 1745, and actually took it after six weeks' siege. Great preparations were then made for an attack upon Canada in the following year; but this came to naught, and in 1748 the war was brought to an end by the Peace of Aix-la-Chapelle.

In the following year the British Government took an important step by founding in Nova Scotia a military settlement, made up of soldiers and sailors who had been disbanded at the close of the war; a settlement which still flourishes under its original name of Halifax. France instantly took the alarm, and began again its former policy of stirring up the Indians of Acadia against our colonists, with the result that there were frequent disputes and petty skirmishes between the French and English. But British traders soon paid off the French in their own coin by penetrating to the Ohio, and stealing away the hearts of the Indians in that quarter. This was a serious menace to the French project of cutting off the British colonists from the interior and confining them to the coast; and accordingly, in 1752, the French Governor of Canada, the Marquis Duquesne [1] sent an expedition to Lake Erie which built two forts: one at Presqu'île, on the site of the present town of Erie, and a second, several leagues to south of it, called Fort Le Boeuf. The British claimed the ground upon which the fort was built as their own; and a young officer of the Virginian Militia, by name George Washington,[2] was sent to request the French garrisons to withdraw. But Duquesne had not secured his communication between the St. Lawrence and the Ohio in order to yield it up at the bidding of an English Governor; and the officers of the two garrisons simply answered that they had orders to take possession of the Ohio, and 'by God they would do it.'

[1] Marquis Duquesne de Menneville, Governor of Canada, 1752–1755. A naval officer who did much to further the efficiency of the Colonist Militia, and to extend the establishment of forts with the object of confining the British within their own colonies, and resisting the threatened tide of invasion.

[2] The future first President of the United States.

With this answer Washington returned to Virginia; and the Governor, Mr. Dinwiddie, without further ado sent a small force to erect a fort at the point where the Monongahela joins the Ohio. The site had been chosen by Washington, and it is that now occupied by the huge city of Pittsburg; but the fort was not destined to be built by the British. On April 17, 1754, before the work had been well begun, a flotilla came paddling down the Alleghany with French troops to the number of 500 men on board. They landed, trained their artillery upon the unfinished stockade, and summoned the British backwoodsmen to surrender. The overwhelming superiority of their numbers left the British no choice. The place was yielded up to the French, who at once demolished the works, and built on the same site a far larger and stronger fort which they named Fort Duquesne.

Governor Dinwiddie, very properly, treated this outrage as a declaration of war; and in the following year two regiments under General Braddock arrived from England to drive the French from the Ohio. Braddock began his advance in May, with some two thousand men; but on July 8 his column was cut to pieces upon the Monongahela River within a day's march of Fort Duquesne, by a mixed force of French and Indians. Subsidiary expeditions directed against Niagara and against Ticonderoga upon Lake Champlain, though not disastrous, were disappointing and achieved no decisive success. Altogether the British campaign of 1755 in America was a total failure.

Meanwhile, at home, affairs were in utter confusion. In spite of the open and undisguised hostilities that had passed between France and England, no declaration of war had yet been made, and the two countries were still nominally at peace; and under the guidance of the Prime Minister, the foolish and incompetent Duke of Newcastle, little or no effort was made towards military preparation.

But Maria Theresa, Queen of Austria, was burning to fight Prussia and recover the province of Silesia. France backed her up; so did Russia. Frederick the Great, King of Prussia, consequently sought an ally, and although on bad terms with his uncle, George II of England, cleverly inveigled him into an alliance by the bait of a guarantee for the province of Hanover of which King George was electoral prince. The British ministers jumped at the alliance for a different reason, viz. to entangle France in European war and thus improve their chance of capturing her possessions in Canada.[1] The House of Commons, too, spurred on by William Pitt the elder, forced the Ministry into action, and in January 1756 orders were given for the raising of ten new battalions, which still form part of the British Army. Finally, during the following month, as the immediate consequence of Braddock's defeat, a Bill was brought in to enable the King to grant commissions to foreigners settled in North America provided—such was the bigotry of Parliament—that they were not members of the Church of Rome;[2] for, since the end of the seventeenth century many provinces—notably Pennsylvania and the two Carolinas—had received numbers of emigrants

[1] The alliance between Prussia and Britain was arranged by the Convention of Westphalia, February 1756. France, Austria, and (soon afterwards) Russia formed a triple alliance in antagonism. So began the Seven Years' War. The pluck and victories of Frederick made England wild with excitement; and in every tavern the English people, whose religious convictions are often hazy, toasted the King of Prussia—an avowed infidel—as the great Protestant hero.

[2] The Naval Service took in Roman Catholics in 1758. In the Army they were admitted—surreptitiously—in 1775, but were not authorised by law until a much later date.

from France, Switzerland, the Palatinate and other parts of Germany. Since the more recent arrivals were to a great extent indifferent as to their flag, it seemed good policy to prevent them serving France in Canada, by inducing them to serve England in America. The Bill was opposed in the House of Commons by William Pitt the elder, on the grounds that British battles should be fought by British troops. But in point of fact these men formed but a small proportion to the colonial recruits, who were either British subjects or had resided for years in America under the protection of the British law. Pitt would seem also to have forgotten that Britain herself had enormously profited at no very distant date, by the immigration of Huguenots and other foreigners; and that the vast and sparsely populated tracts of America would *a fortiori* develop the more rapidly by the introduction of settlers accustomed to military discipline. Wiser counsels happily prevailed. The Bill was passed and the four new battalions were voted without discussion, the Act merely stipulating that the foreign soldiers should not serve out of America, and that no foreign officer should hold rank higher than that of lieutenant-colonel. Major-General the Earl of Loudoun, who had been appointed to succeed Braddock as supreme commander in America, was created Colonel-in-Chief—a title granted to the head of no other regiment, although many had more than one battalion—and, that the King's favour should not be wanting to encourage the new corps, it received, as already noticed, the title of the 'Royal American Regiment.' This was a special compliment, for regiments at this time were still known chiefly by their Colonel's name, unless they possessed a permanent title.[1]

[1] Numbers were seldom or never quoted in official correspondence, though they were employed to denote precedence in rank.

There were also reasons of State behind the title. The word 'American' associated the Regiment with the colonies whence its recruits were mainly to come. The word 'Royal' bound it, *ab initio*, to the British Crown.

For the sake of the French, Swiss, Tyrolese, and German settlers who were expected to enlist in the Regiment commissions were sanctioned for fifty foreign officers. Provision was also made for twenty foreign Engineer officers belonging to the Regiment, but exclusive of its establishment.

Colonels Stanwix, Dusseaux, Jeffereys, and Jacques Prevost were appointed Colonels-Commandant of the 1st, 2nd, 3rd, and 4th Battalions respectively. The Lieutenant-Colonels were Henri Bouquet, Frederick Haldimand, Russell Chapman, and Sir John St. Clair; of whom the last named had been wounded in Braddock's action.

To emphasise the fact that the Regiment would be chiefly employed in bush warfare, the uniforms were made devoid of lace—a step, even though but a short one, in the right direction.

The new corps was first known as the 62nd, but the disbandment, in America, of two regiments captured by the French at Oswego, caused it almost immediately to become the 60th.

The *London Gazettes*, Nos. 9566 (March 16–20, 1756) and 9569 (March 27–30, 1756), give a list of appointments to the Regiment as follows:

The King has been pleased to constitute and appoint the following Lord and Gentlemen to be Officers in the 62nd or Royal American Regiment of Foot to be forthwith raised for His Majesty's Service in North America :—

Colonel-in-Chief

The Right Honourable John Earl of Loudoun,[1] Major-General of His Majesty's Forces.

Colonels Commandant

John Stanwix, Esq.
James Prevost, Esq.[2]
Joseph Dusseaux, Esq.

Lieutenant-Colonels

Henry Bouquet, Esq.
Frederick Haldimann,[3] Esq.
Russel Chapman, Esq.
Sir John St. Clair, Bart.

Majors

John Young, Esq.
James Robertson, Esq.
John Rutherford, Esq.
Augustin Prevost, Esq.

Captains of Companies

John Tullikins, Esq.
Thomas Oswald, Esq.
Rodolf Faesch, Esq.
Frederick Porter, Esq.
—— Munster,[4] Esq.
Walter Rutherford, Esq.
—— Wettsteen, Esq.
Ralph Harding, Esq.
—— Chambrier, Esq.
Jeremiah Stanton, Esq.
—— Knielling,[5] Esq.
Richard Mather, Esq.
Gustavus Wetterstroom, Esq.
Harry Charteris, Esq.
Paul Castleman, Esq.
—— Stiener, Esq.[6]
Francis Lander, Esq.
—— Rollaz, Esq.
John Innis, Esq.
—— Schrader, Esq.
Gavin Cochran, Esq.
Joseph Prince, Esq.
Marcus Prevost, Esq.[7]
Thomas Stanwix, Esq.
Alexander Harbord, Esq.
Abraham Bosomworth, Esq.
John Faesch, Esq.

[1] The first Colonel-in-Chief of our Regiment was a man of no genius, yet filled several important posts in the British Army without discredit. In the '45' he acted as Adjutant-General. After quitting America, he was employed as Second-in-Command of a British force in Portugal. Died 1782, aged 77.

[2] Afterwards Governor of Antigua.

[3] The German method of spelling Haldimand.

[4] Dietrich Herbert, Freiherr v. Munster.

[5] Spelt sometimes Gnielling.

[6] Lewis Stenier.

[7] Jean Marc Prevost.

Captain-Lieutenants

—— Konn, Esq.
John Dalrymple, Esq.
Stephen Gually, Esq.
Edward Comberbach, Esq.

Lieutenants

Captain Andrew Nisbit, from Half Pay, Irish (Bruce's).
Lieutenant George Macadam [1] ditto.
Lieutenant Charles Crookshanks, from Half Pay (Leighton's).
Lieutenant George Brereton, from Half Pay (Bragg's).
Lieutenant Francis Pringle, from Scotch Dutch.
Lieutenant Robert Brigstocke, from Half Pay (Richbell's).
Lieutenant Allan [2] Macbean, from Half Pay (Lord John Murray's Additionals).
Lieutenant Donald Campbell (Dutch).
Lieutenant Newsham Piers, from Half Pay (Barrell's).
Second Lieutenant —— Longdon, from Half Pay, Irish (Pole's).
Bazil Dunbar, Dutch Service.
Robert Drew, from Half Pay, Irish (Pole's).
—— Ourrey,[3] Fort Adjutant, Jersey.
Second Lieutenant Richard Wynne, from Half Pay, Irish.
Second Lieutenant Ebenezer Warren, from Half Pay, Irish (Otway's).
William Baillie, from Half Pay, Dutch.
Second Lieutenant John Swift, ditto, Irish (Pole's).
Second Lieutenant John [4] Cooke, from Half Pay, Irish (Bruce's).
Ensign Henry Symcocks, from Half Pay, Irish (Pole's).
Ensign Charles Wellington, from Half Pay, Irish (Irwin's).
Ensign John Sealy, from Half Pay, Irish (Pole's).
Ensign James Campbell, from Loudoun's.
Second Lieutenant Simon Fraser, Dutch Service.
Ensign George Fullerton, from Loudoun's.
Ensign William Stuart, from Anstruther's.
Second Lieutenant Alexander Campbell, from Half Pay, Dutch.
Ensign —— Ray, from Rich's.
Second Lieutenant George Turnbull, from Half Pay, Dutch.
Ensign William Abercromby, from Abercrombie's.
Second Lieutenant —— Ouchterlony, from Half Pay, Dutch.
Quarter Master William Hazlewood, from Herbert's.
Ensign —— Brown, from Half Pay, Dutch.

[1] *Gilbert* Macadam in *Army List*.
[2] In *Army List*, Alexander.
[3] Louis Ourry.
[4] William, in *Army List*.

Adjutant M'Alpin, from Half Pay (Oglethorpe's).
Donald Forbes, Dutch.
Engineer Henry Gordon.
Lieutenant —— Mackay, Dutch.
Engineer Thomas Basset.
Sergeant John Elrington,[1] from Jorden's.

Ensigns.

Sergeant Major Allan, from Leighton's.
Sergeant Major Barnsley, from Loudoun's.
Sergeant M'Intosh, from Cope's.
Thomas Campbell.
R. Phillips, from the Horse Guards.
Samuel Mackay, from Dutch Guards.
Francis Mackay, ditto.
—— Archibold.
James Munro.
William Ridge.
William Hay.
—— Shaw.
Thomas Meredith.
—— Kerr.[2]
Walter Kennedy.
Michael Davies.
William Potts.
—— Watson.
—— Jones.
Nicholas Sutherland.
William Ryder.
Abraham Hart.
Robert Campbell.
James Crofton.
Townshend Guy.
James Jeffries.
Francis Hutchinson.
James Herring.
—— Jenkins.
—— Ralph.[3]
John Nuterville.
Edward Obrien.

The *Gazette* does not include all the Engineer officers appointed to the Regiment: a full list of the original regimental officers as shown in the *Army List* in given in Appendix VI.

NOTE.—At this period the highest regimental rank for which pay could be drawn was that of captain. Every colonel commandant, lieutenant-colonel, and major therefore had a company. The company of the first-named was actually commanded by the senior subaltern,

[1] George Etherington, in *Army List.*
[2] Carre, in *Army List.*
[3] Ralfe, in *Army List.*

under the title of Captain-Lieutenant, which term is to be found in the *Army List* up to 1803. For instance, in the *London Gazette*, No. 9566, containing part of the above-named appointments is a notice that 'The King has been pleased to constitute and appoint James Abercrombie, Esq., Major General of His Majesty's Forces'—subsequently Colonel-in-Chief of the Royal Americans—'to be Colonel of the Regiment of Foot, late under the command of Colonel Ellison deceased; and likewise to be Captain of a Company in the said regiment.'

JOHN, EARL OF LOUDOUN

(From the portrait by Allan Ramsay)

(From the portrait by Allan Ramsay)

CHAPTER I

PROGRESS OF THE ROYAL AMERICAN REGIMENT—STRATEGICAL FEATURES OF THE AMERICAN COLONIES—LIEUT.-COLONELS BOUQUET AND HALDIMAND—MOVEMENTS OF THE REGIMENT—LOCAL DIFFICULTIES—LOSS OF FORT WILLIAM HENRY—MUSTER ROLLS OF THE THIRD AND FOURTH BATTALIONS

CHRISTMAS DAY, 1755, being the date of Lord Loudoun's commission as Colonel-in-Chief, is usually considered to be the birthday of the Regiment. The commissions of Major John Young and Captain John Tullikens bore the same date, while the bulk of the remainder were signed a few days later,[1] but the actual order to raise the Regiment was not issued until March 4, 1756, when its establishment was fixed at four battalions, all of ten companies. Each company consisted of four sergeants, four corporals, two drummers, and a hundred 'private men,' making for the Regiment a total strength of 4400 N.C.O.s and men.

On March 13, instructions were sent to Colonel Daniel Webb, Commander-in-Chief in America pending Lord

[1] By the end of February the List of Officers was complete; two-thirds bore unmistakably British names, the Scots being predominant. Certain inducements had been held out to them to accept service in America, among which was the promise of grants of land. A Field Officer was to be entitled to 5000 acres, a Captain to 3000, a Subaltern to 2000; an N.C.O. was to receive 200 acres, a Private 50. At that period these grants may have appeared but poor recompense; in the twentieth century the imagination is staggered at the probable present value of the lands promised and no doubt largely given to the original members of our Regiment.

Loudoun's arrival, that the province of Pennsylvania was to be reserved as a recruiting ground for the Royal Americans. Now in Pennsylvania the Society of Friends, more commonly known as Quakers, was supreme. They must have received the news with mixed feelings. Since their religious principles debarred them from military service, it must have been satisfactory to have among them a body of men who could defend them from the French and Indians; on the other hand, it must have gone against the conscience of men who would only vote money to buy gunpowder under the guise of 'corn and other grain,' to see the recruiting sergeant enlisting men under their very noses for the wickedness of warfare.

No time was lost in placing the newly gazetted officers under orders to proceed to America. Among the first to embark were Lieut.-Colonels Bouquet and Haldimand, Major Prevost, Captain J. M. Prevost, and Lieutenant Meyer. Bouquet and Haldimand reached America on June 15; Lord Loudoun did not start until May 20, and took exactly two months to reach New York.

Among the towns and more thickly populated districts of Pennsylvania recruiting at first seems to have made slow progress. A large portion of the labouring class were in a condition not very much higher than slavery. Nominally they were free men who had merely engaged not to quit the employment of their masters for a specified term of years; but in point of fact numbers had been kidnapped by crimps who practically sold them to Americans at so much a head. To join the Royal American Regiment was to these men to recover their liberty; but in deference to their masters enlistment was at first forbidden. The Quakers having obtained this concession at once declared every recruit to be an indentured servant. This was too much: the prohibition was withdrawn; all men, whether indentured or not, could be enlisted; and it was not unnatural that the masters complained loudly and that a bitter feeling arose between the civilian element and the soldiery.

But in the western settlements of the American plantations, far from civilisation and luxury, abode a large colony of backwoodsmen, partly English, partly Swiss, Tyrolese, and Germans—a strong and hardy race accustomed to the country and inured to the climate; from their conditions and habits of life well adapted to oppose the enemy, whether French or Red Indians; and of the very material to carry on frontier warfare in such a country as is described. It was of these that the Regiment was at the outset largely composed; and it is interesting to note that not only were they accustomed to the use of the rifle, but that it was by these Swiss and Tyrolese that the weapon had been introduced into America.

The officers appointed to the new corps seem to have been a singularly efficient body, judging from the number whose fame still survives on the far side of the Atlantic. Stanwix, Jacques Prevost, Haviland, Colonels-Commandant; Bouquet, Haldimand, John St. Clair, Lieutenant-Colonels; Young, Augustin Prevost, Marc Prevost, Baron Dietrich Herbert Münster, Hollandt, Ourry, Ecuyier, Etherington, Ratzer, Des Barres, Fuser, and others—all to a greater or less extent made their mark upon American or Canadian history. So did Lord Howe, Lawrence, Monckton, Bradstreet, and Arthur St. Clair, appointed to the Royal Americans in 1757.

But what specially distinguished the new regiment was the fact that among the seniors were several who had either served in the French army, or were at all events fully acquainted with its system. British regiments had been hitherto brought up on the rigid Prussian model, and drilled as unreasoning machines. Colonels Bouquet and Haldimand cultivated the intelligence of each individual member of the corps. This in after days was the system of Sir John Moore, on which he trained the famous 'Light Division.' There is a story that Moore declared he learnt it from the 60th in which he served for a short time. The story may be apocryphal, yet not impossibly has some vestige of foundation in fact. It is the system which—particularly since the war in South Africa—has been inculcated in the army at large; but in the 60th it was innate, and has been practised during the whole of its history.

At the period of which we are speaking the American colonies formed little more than a strip along the shores of the Atlantic, shut in on the west by the slopes of the Alleghany Mountains. Beyond the watershed of this range adventurous spirits indeed dwelt; but these back-settlements were sparsely occupied, and the country was covered with forest and morass, almost impenetrable, hardly explored. Mountain and valley were intersected by deep creeks and rivers; bridges and roads were almost unknown, and the sole means of traversing the woods was by the track or pathway known only to the Indians or to a few of the settlers. It need hardly be pointed out that in a country of this kind the exact formation and precise movement of regular troops trained under the Prussian system were impossible. Better adapted to the guerilla warfare necessitated by the *terrain* and the enemy, were the provincials and friendly Indians, imbued as they were with the hunter's instinct, dexterous with the rifle and hatchet, and skilful in the necessary tactics. By the adoption of Indian dress and close study of Indian methods Bouquet resolved that his regiment should be trained to meet the Red-skins on equal terms.

As to the strategical situation, the main base of the French was Quebec, their one port of communication with France, while Montreal formed their advanced base for all operations inland. Louisbourg was a subsidiary naval one, intended for raids upon the coast of New England and Newfoundland; but, being dependent upon the sea for all supplies, was, as already pointed out, an embarrassment rather than a help. The communication of Montreal with the back-country lay south-westward along the St. Lawrence to Lake Ontario, through the intermediate fortified posts of Chambly, Oswegatchie or La Galette, and Fort Frontenac—the last lying at the head of the lake itself. Thence, turning westward across Lake Ontario, the next post was Niagara, upon the river that connects Ontario with Lake Erie. Here was a strong fort to cover the carrying place necessitated by the falls of the river. Upon the south-eastern shore of Lake Erie was Fort Presqu'île, with a small fort, Le Boeuf, immediately to the south of it upon La Rivière des Boeufs (or, as the British called it, French Creek), which formed the northern gate of the French settlements on the Ohio. To the south of Fort Le Boeuf, at the junction of French Creek with the Alleghany River, was the fort of Venango; and south of Venango, still upon the united rivers, stood the famous Fort Duquesne, the eastern gate of the French back-settlements on the Ohio and the Mississippi.

But the chain of the great lakes afforded another passage to those same back-settlements which was equally secured by fortified posts. The channel from Lake Erie to Lake Huron was dominated by the fort of Detroit; that from Lake Huron to Lake Superior and on to the head-waters of the Mississippi, by Fort Ste. Marie; that from Lake Huron to Lake Michigan, by the Fort of Michilimackinac or Mackinaw. On a river at the south-western corner of Lake Erie, Fort Miamis guarded the entrance to the eastern head of the Illinois. Far to the south the posts of Fort de Chartres on the Illinois and of Fort François on the Ohio asserted the French sovereignty over the vast territory in the hinterland of the British settlements as far as New Orleans.

But these chains of posts were long and weak. A glance at the map will show that the capture of Niagara would cut off the whole of the French back-settlements both on the Ohio and the Illinois; that the capture of Quebec, so long as Great Britain was supreme on the ocean, would place their colonies in the position of a beleaguered garrison; but that the heart of New France was Montreal.

Turning to what are now the United States, but at that time were our own colonies, it may be said that while innumerable ports were open to the British, their base was, and always had been, New York, communicating by the River Hudson with their first advanced post at Albany, about one hundred and fifty miles to the north. From Albany there was a choice of two routes for an advance upon Montreal. The first lay to westward up the Mohawk River to the Great Carrying Place, and thence overland to Oswego upon Lake Ontario. This was the natural line for an attack upon Niagara. The second followed the Hudson northward from Albany to Lake Champlain which could be approached either by way of Lake George or by a river called Wood Creek, running parallel to it on the east. This was the true line for an attack from Montreal upon New York, or from New York upon Montreal. The British had closed their end of the great waterway by Fort William Henry at the southern end of Lake George, by Fort Anne upon Wood Creek, and by Fort Edward at the spot where, to the south of Fort William Henry, the line of advance left the Hudson. The French on the other hand had barred the southern end of Lake Champlain by the fort of Ticonderoga, or Carillon, built upon the headland where the waters of Lake George and Wood Creek unite, and by a second fort at Crown Point, a headland eighteen miles north of Ticonderoga. North of Lake Champlain the access to the St. Lawrence by the River Richelieu was defended by the three posts of Isle aux Noix, St. John's and Fort Lévis; and by Fort Sorel at the junction of the Richelieu with the St. Lawrence itself.

For operations against Fort Duquesne better and more convenient bases than New York were to be found both at Philadelphia and on the James River; but Philadelphia was preferable, both because the line was shorter, and because Pennsylvania was far better provided with forage and the means of transport. The route from Philadelphia lay due westward by way of Raystown;[1] that from Virginia involved first the despatch of transports by sea from Williamsburg up the Potomac to Alexandria, and then a tedious advance, still up the river, to the advanced base at Fort Cumberland. This last was the line selected by Braddock under instructions from England and by the advice of Washington; but since the disastrous defeat of his column nothing more had been attempted in that quarter.

In 1755 attempts had been made by British Colonial troops both upon Niagara and Crown Point. The column destined for Niagara was, however, compelled by want of supplies to fall back, leaving seven hundred men to hold Oswego; while that which moved upon Crown Point succeeded indeed in beating off a French attack, but could not do more than build the two forts William Henry and Edward, and leave a considerable garrison in both of them. In the summer of 1756 these two columns were to have

[1] Sometimes spelt Reastown.

advanced again, but the colonial troops had become so much demoralised by enormous losses from sickness, due to the filthy state of their camp, that Lord Loudoun resolved to abandon the attack upon Niagara, and to turn his whole strength against Lake Champlain. Before, however, he could do so, the Marquis de Montcalm, an officer but recently arrived in Canada as Commander-in-Chief of the French forces, swooped down upon Oswego in overwhelming strength, and in three days compelled it to surrender. Reinforcements were on their way to the place at the time, but Colonel Webb who was in command of the district made no attempt to recover it. Smitten with panic, he burned the fort which had been constructed at the Great Carrying Place, and retreated precipitately down the Mohawk. Montcalm then established himself at Ticonderoga in a position unassailable by the small force at Lord Loudoun's disposal, and the whole of the operations came to a standstill.

It was caustically remarked of Lord Loudoun that, like the figure of St. George on a tavern sign, he was always galloping yet never advancing; but in the present instance there was perhaps no alternative for him but to place the troops in winter quarters; and his first reference to his new regiment appears to be made during the process of this distribution. Writing on October 3, he says: 'Three battalions of the Royal Americans are still to raise, desertion is beyond all bounds; I have lost 30 deserters of the Royal Americans since they came to Albany.' This was no new trouble in America; it was the habit of the colonists to implore military aid from England, but when the troops arrived they would refuse the men quarters, and gain a little cheap labour by taking advantage of their misery to induce them to desert. However this first contingent of the Royal Americans was preserved, despite the loss of the thirty men, and was temporarily divided into two battalions ; the first under Lieut.-Colonel Bouquet and Major Young, the second under Lieut.-Colonel Haldimand and Major Robertson.

The characteristics and careers of the first two lieutenant-colonels of our Regiment cannot fail to excite curiosity and interest, and the lives of both will amply repay careful study.

Henri Bouquet—for wider details of whose life Sir Edward Hutton's admirable monograph should be consulted [1]—was born in 1719, at Rolle, near the northern shore of the Lake of Geneva. Of his family nothing is known ; he seems to have had few relations in Switzerland, but a gentleman named Buckey, living at the present date (1913) in the United States, claims to be descended from the colonel's brother, who no doubt followed Henri to America. According to his portrait Bouquet's features are not lacking in refinement, and ability is shown in every line ; the eye gives evidence of the alertness and vivacity essential to a Rifleman ; the mouth indicates resolution ; but there is no asperity in his countenance, which is perfectly good-humoured. The general impression gained is that of a man who thoroughly knew his own mind and meant to carry out his views, yet without riding roughshod over his subordinates.

Although French was Bouquet's native tongue, he spoke and wrote English well, and was certainly a man of exceptional reading and education. At the age of seventeen he entered the Dutch service as a cadet in the regiment of Constant, receiving a commission as ensign two years later. Then we are told that he served with distinction in the Sardinian army against France in Northern Italy during the War of the Austrian Succession. In 1748 at

[1] *Colonel Henry Bouquet* (Warren & Son, Ltd. : Winchester, 1911).

the peace of Aix-la-Chapelle Bouquet re-entered the Dutch service, and became captain and lieutenant-colonel in the Swiss Guard, now raised by the Prince of Orange, whose attention he had attracted during the campaign. His first employment was in the responsible duty of occupying the forts evacuated by the French under the terms of the Treaty. Bouquet's subsequent career will be traced in the following pages. The mass of literary matter which he left behind at his death gives evidence of intimate knowledge of all branches of his profession; his regulations—the dress, equipment, training, and tactics required—for service against a savage foe may be studied with advantage at the present day;[1] and not less in personal skill as a scout than in advanced views and lofty ideals, Henri Bouquet has claims to be considered the father of British Riflemen.

Frederick Haldimand, also a Swiss, was a year older than Bouquet, and spent his boyhood at Yverdun in the Canton of Vaud. The picture of him late in life by Lemuel Abbot,[2] gives evidence of breeding and refinement of feature, combined with the air of a soldier. Haldimand, another soldier of fortune, who is believed to have begun his career in the Sardinian service, served at Molwitz, under Frederick the Great, whom he ever afterwards spoke of as 'my old master.' Later on, in the Dutch service, he was thrown with Bouquet—his intimate friend and the object of his unbounded admiration.

In North America Haldimand served for thirty years, becoming Governor of Canada; in which capacity, despite criticism, he certainly achieved a large measure of success. The lawyer could not understand that he could dispense

[1] *Vide* p. 160 and Appendix IV.

[2] The picture, at present in possession of the writer, may be that attributed at first to Reynolds, with whose style Abbot's work was often compared. If it be so, this portrait was stated by Queen Caroline to be very like Haldimand.

with precedents; the politician, that he was not saturated with party spirit; the minister of religion, that he could uphold a Church to which he did not belong. But Haldimand carried out his resolve to maintain the law with impartiality, and Canada owes the General a debt of gratitude. He was made a Knight of the Bath, and on his return to England in 1786 was treated with confidence and friendship by the King, who on one occasion said to him, 'You are a soldier, and are always right.' In 1791 he went back to die at his old home at Yverdun. A tablet in his honour, erected before his death, is still to be seen in Henry VII's Chapel at Westminster Abbey.

'He had helped,' says Bradley, 'to recruit the Royal American Regiment among Oglethorpe's Highlanders of Georgia, the Germans and Scotch, Irish of Pennsylvania, and the indentured servants, poor whites and Huguenots of the two Carolinas. . . . His military services were considerable, and above all he was Governor of Canada during the revolutionary war from 1778 to 1784—a sufficiently critical and conspicuous post at that time, which he admirably filled.'

The 'Makers of Canada' series contains an excellent monograph on Haldimand.

The two battalions did not long remain together, for on September 28 one of them was moved a day's march to Fort Edward, while the other remained at Saratoga. Our next notice of the Regiment is a list of about 250 non-commissioned officers and privates who were turned over to the Royal Americans at Boston on March 7, 1757, and were evidently the remains of two regiments, once numbered the 51st and the 52nd, which were disbanded at this very time. Next in order comes a letter, dated April 25, 1757, from a private, in the following words: 'Sir, I beg the favour of you if you could be pleased to beg of my Lord

[1] *The Fight with France*, p. 343.

to grant me my discharge, and oblige your humble servant, Fr. Weissenfels.' The reason for this sudden petition does not appear; but it is tolerably safe to conjecture that Lord Loudoun, who was much incensed at the loss of so many men by desertion, was not very willing to grant discharges.

About this time Colonel Bouquet was sent back to Philadelphia, whither Brigadier-General John Stanwix, Colonel-Commandant of the 1st Battalion of the Royal Americans, had been despatched in the previous November to conciliate the magnates of Pennsylvania. Whether he succeeded or not is unknown, but Bouquet on arrival had every cause for complaint. Despite their name, the Philadelphians had no compunction in allowing the troops to die of cold and disease. Bouquet's tact in dealing with civil authorities is proverbial, and when he stated that his men were cruelly and barbarously treated, we may feel sure that he was not overstating the case. The number to be provided for was about 550, and the public houses were inadequate for the purpose. Bouquet appealed to Mr. Denny, the Governor, who gave him a warrant for the quartering of men in private houses. The Provincial Assembly flared up; a quarrel between it and the Governor seemed inevitable, but at the last moment the Assembly gave way.

Five companies of the battalion remained under Stanwix in Philadelphia; the other wing (commanded by Bouquet), weak in numbers, sailed thence on May 16, 1757, by Loudoun's orders for Charleston, to form part of a force designed for the defence of North Carolina, South Carolina, and Georgia. The full number of the detachment was 2000 men, all of them, with the exception of the Royal Americans, provincial troops; but of this strength the greater portion existed only on paper, and was to be enlisted by some means or another within the three provinces. Small-pox had been rife among Bouquet's companies at Philadelphia. It broke out again on the voyage, and continued with them after their landing; yet recruiting parties were at once sent out with power to offer a bounty of three pounds, while a reward of one dollar was offered to the non-commissioned officers and drummers for every man enlisted. The numbers of the battalion did not, however, increase quickly. 'The lawyers, justices of the peace and people at large are all against us,' wrote Bouquet. . . . 'The people are pleased to have soldiers to protect their plantations, but will feel no inconvenience for them, making little difference between the soldier and the negro.' Under these circumstances it is not surprising that recruits were difficult to find, and of poor quality when found. Two officers in South Carolina, with great difficulty and expense, raised twelve men in two months. As to the original men brought from Philadelphia Bouquet was little more sanguine. 'We have got a number of such drunken dirty fellows,' he wrote, 'that we shall never make anything of.' Meanwhile the provincial authorities refused to give the men decent quarters; refused even to provide them with straw; and the natural result was much desertion. Charleston was moreover in an extremely unhealthy state; the air, according to Bouquet's account, so infected that the very horses died; it is therefore no matter for wonder that death, not less than desertion, claimed a share of the new battalion. The camp was full of water. On July 29, fear of infection having passed away, four empty houses were handed over to Colonel Bouquet, but they were unfurnished and in bad repair. The men slept on the bare floors. The mortality was still great. Writing to a friend on September 27, Bouquet says: 'Our men die very fast, and we have lost more in one month than in the whole winter at Philadelphia.' Not until the 21st of that month had billets been provided for 160 men. At length, on November 14, the Assembly met and voted a blanket to every two men. None of the usual allowances were made to field officers or captains, but the munificent sum of one shilling a day was granted to the subalterns. After five months of Carolina, Bouquet had but 300 men fit for duty.

By October he was in despair of completing even his five companies to the full strength ; and then was added to him a new anxiety, over and above all other troubles. There were symptoms of an epidemic of matrimony among the officers. 'Captain Lander has married one of our beauties,' he wrote ; 'I hope it will spread no further, as there is no great temptation.' Fortunately the plague seems to have been stayed ; though it is hardly surprising to find that Captain Lander, within a twelvemonth, desired to sell his commission. Meanwhile, under Bouquet's excellent training the men improved rapidly ; for which improvement Bouquet, with the generosity of a great mind, gave all the credit to his second in command, Major Young, whom he described as 'the best major I know in the world.' But at length the hope of completing the half battalion to its establishment in Carolina seems to have been abandoned. In February 1758 the Assembly grudgingly supplied the troops with better quarters, but it was too late. During the first week of March, Bouquet and his five companies were recalled to New York. Here the men were quartered in barracks, but the officers were left out in the cold. Lord Loudoun demanded provision for them in no uncertain tone. 'G—— d——n my blood,' shouted he at the Mayor who came to remonstrate, 'if you do not billet my officers upon free quarters this day,

I'll order here all the troops in North America and billet them myself upon this city.' To enforce his words he at once sent six soldiers to the house of Delancey, a most influential citizen. Delancey swore at them; whereupon Loudoun, despite his own surpassing profanity, shocked, like a true Scot, at bad language in others, sent six more! The townspeople then raised a subscription and provided the quarters. At New York recruits from the outlying country districts by degrees raised Bouquet's battalion to the strength of its establishment.

The following quaint 'Beating Order' gives evidence of Lord Loudoun's energy in recruiting:

BEATING ORDER

NOVEMBER 16TH, 1757

Albany

BY His Excellancey John Earl of Loudoun Lord Machline and Tarrinzean &ca. &ca. &ca., one of the Sixteen Peers of Scotland Governor and Captain-General of Virginia and Vice-Admiral of the Same; Colonel of the Thirtieth Regiment of Foot, Colonel in Chief of the Royal American Regiment; Major-General, and Commander-in-Chief of all His Majesty's Forces raised or to be raised in North America.

These are to authorize you by beat of Drum or otherwise, to Raise in any County or Part of the Colonies on this Continent, so many voluntiers as are or shall be Wanting to Recruit and fill up the Respective Companies of the First Battalion of his Majesty's Sixtieth Regiment of Foot, under Your Command and all Magistrates Justices of the Peace, Constables and other Civil officers are required to be assisting unto you in Providing Quarters Impressing Carriages &c. as there shall be Occasion.

Given under my hand at Albany this 16th Day of November 1757.

(Sign'd) LOUDOUN.

By his Excellency's Command
(Sign'd) J. APPY,
(N.B. The Judge Advocate).

The progress of the battalions, other than the 1st, is extremely difficult to trace. Haldimand's, which was numbered the 2nd, seems to have taken root first at New York, and to have raised most of its numbers there. In September 1757 the Colonel inspected the camp of the battalion, then at Carlisle on the right bank of the Susquehanna, and found it short of ammunition and stores. The 3rd appears to have been quartered during the earlier part of 1757 at Albany; its commander being nominally Lieut.-Colonel Chapman, but really, so far as can be ascertained, Lord Howe, Colonel-Commandant of the battalion, one of the best officers in the army, and, like Bouquet, one who held advanced views on the dress, training and equipment of light infantry. The 4th Battalion was with the 2nd at New York, but in October had moved to Albany.

It took a long time to begin the campaign of 1757. Lord Loudoun, who had received promises of a reinforcement of seven battalions, wished to attack Louisbourg; but these troops did not arrive until July; nor, when they at last made their appearance, was there any fleet to escort them. Risking, however, the hazard of being destroyed at sea, he embarked his army, including the 2nd and 4th Battalions of the 60th, and made his way safely to Halifax. There, on July 9, Loudoun was joined by Admiral Holburne's squadron, which proved to be so much weaker than the French that he abandoned the entire expedition and sent back to New York the troops that he had taken therefrom.

A part of the 3rd Battalion had in the meantime moved up from Albany to Fort Edward, whence, on August 2, a detachment 200 strong under John Young (now lieutenant-colonel) was despatched by Colonel Webb to form part of the garrison of Fort William Henry under Colonel Monro of the 35th. Next day this fort, with its wooden parapet, and garrison of 2200 men, one-third only being British regular troops, was besieged by a sudden advance of 6000 French and 1600 Indians under the Marquis de Montcalm.[1] Webb, from Fort Edward, fourteen miles distant, had reinforced Monro with 1000 men and had only 1200 remaining. Rightly or wrongly he made no further attempt to aid him. The parapet was quickly battered down by the French, and the British guns were silenced. More than 300 of the garrison had been killed and wounded and small-pox was raging in the fort. On August 9 Colonel Young, who had been wounded, was despatched by Monro to negotiate terms of surrender. It was stipulated that the British should march to Fort Edward in safety with their small arms and one field-gun. The fort then capitulated. Montcalm, assisted by the regular officers of the French army, of which no less than six battalions were present, made every personal attempt honourably to fulfil the treaty. But why did he not get his regular troops under arms to enforce it? The fact is that the French did not dare to offend their Indian allies, and the Canadian militia thirsted for blood. A terrible scene ensued. Undeterred by their white comrades, the Indians broke in upon the prisoners and massacred about 80 soldiers, besides women and children. The story of the massacre of Fort William Henry is told by Fenimore Cooper in his 'Last of the Mohicans.'[2] This atrocity, like that of Cawnpore a century later, created a furore which took many years to subside, and is not forgotten at

[1] Six companies of the 35th were present. Another return estimates the garrison at 2450, in which case the proportion of British regular troops was only one-fourth of the whole.

[2] Fenimore Cooper seems to have been son-in-law to Captain J. Delancey, of the Royal American Regiment, who may have been present at the massacre, for he belonged to the 3rd Battalion.

the present day. Its immediate result was that 'Remember William Henry' became the stern reply to many a piteous appeal for quarter by Frenchman or Canadian. By death, disease, wounds, and probably massacre the 3rd Battalion seems to have suffered a loss of about 80 men in this affair. It was not easy to replace them. The battalion was mustered at Fort Edward in October and found to number 30 officers and 413 of all other ranks. Eleven officers and 325 of other ranks are described as 'absent': these numbers signifying probably the difference between the actual strength and the establishment.

The details of this muster, as supplied by W. O. Muster Rolls, may be of interest.

60TH FOOT. ROYAL AMERICAN REGIMENT. 3RD BATTALION.

Muster taken at Fort Edward on the 24th October 1757, for 183 days, from the 25th April 1757, to the 24th October following, both days inclusive.

Colonel-in-Chief The Earl of Loudoun.

Commanded by Viscount Howe.

Major Augustin Prevost's Company (Major and Captain).

Effective	Major and Capt.	Lieuts.	Ensigns	Sergeants	Corporals	Drummers	Private men.
Present	1	2	1	3	3	1	30
Absent	–	–	–	1	1	1	30
Total	1	2	1	4	4	2	60

Captain James Delancey's Company.

	Captains	Lieuts.	Ensigns	Sergeants	Corporals	Drummers	Private men.
Present	1	1	–	2	3	2	36
Absent	–	1	1	2	1	–	39
Total	1	2	1	4	4	2	75

Captain Jeremiah Stanton's Company.

Present	1	1	–	3	2	1	29
Absent	–	1	1	1	2	1	39
Total	1	2	1	4	4	2	68

Captain Gmeling's[1] *Company.*

	Captains	Lieuts.	Ensigns	Sergeants	Corporals	Drummers	Private men.
Present	–	1	1	2	4	–	33
Absent	1	1	–	2	–	2	35
Total	1	2	1	4	4	2	68

Captain William Littler's Company.

	Captains	Lieuts.	Ensigns	Sergeants	Corporals	Drummers	Private men.
Present	1	1	1	4	1	1	35
Absent	–	1	–	–	3	–	35
Total	1	2	1	4	4	1	70

Captain Harry Charteris' Company.

	Captains	Lieuts.	Ensigns	Sergeants	Corporals	Drummers	Private men.
Present	1	1	1	3	3	2	33
Absent	–	1	–	1	1	–	39
Total	1	2	1	4	4	2	72

Captain Rudolphus Faesch's Company.

	Captains	Lieuts.	Ensigns	Sergeants	Corporals	Drummers	Private men.
Present	–	2	1	1	3	2	37
Absent	1	–	–	3	1	–	32
Total	1	2	1	4	4	2	69

Captain Frederick Porter's Company.

	Captains	Lieuts.	Ensigns	Sergeants	Corporals	Drummers	Private men.
Present	1	3	–	4	4	2	68
Absent	–	–	–	–	–	–	20
Total	1	3	0	4	4	2	88

Captain-Lieutenant Edward Comberbatch's Company.

	Col. & Capt.	Capt.-Lieut.	Lieut.	Ensign	Sergeants	Corporals	Drummers	Private men.
Present	1	1	–	1	3	3	1	49
Absent	–	–	1	–	1	1	1	30
Total	1	1	1	1	4	4	2	79

In this last (which is usually designated the 'Colonel's') company were also included the Chaplain, Adjutant, Quartermaster, Surgeon, and Surgeon's mate.

[1] In *Army List* this officer is called Knielling or Gnielling. In the Muster Rolls his Christian name is stated in one place as Gottlieb, in another later on as George Adam!

From Fort Edward the battalion moved during the winter to Dartmouth, Nova Scotia, where, on May 6, 1758, it was again mustered for the period October 25, 1757, to April 24, 1758, when it numbered 35 Officers and 961 N.C.O.s and men 'present': with 9 Officers and 50 N.C.O.s and men 'absent.' The rapid numerical increase seems to give evidence of the popularity of the regiment.

For the half-year ending October 24, 1757, the Muster Rolls give the following figures for the 4th Battalion at the camp near Albany:

		Officers	N.C.O.s & men.
Colonel James Prevost's Company	Present	7	42
	Absent	2	46
	Total	9	88
Captain Alexander Harbord's Company	Present	3	34
	Absent	1	49
	Total	4	83
Captain Lewis Steiner's Company	Present	3	37
	Absent	1	42
	Total	4	79
Captain John Innes' Company	Present	3	24
	Absent	1	55
	Total	4	79
Captain Marcus Prevost's Company	Present	2	39
	Absent	2	42
	Total	4	81
Captain Walter Rutherford's Company	Present	3	60
	Absent	–	31
	Total	3	91

		Officers	N.C.O.s & men.
Captain Sam. Willyamoz' Company . .	Present	1	23
	Absent	3	55
	Total	4	78
Captain George Du fez' Company . .	Present	3	35
	Absent	1	49
	Total	4	84
Major John Rutherford's Company . .	Present	3	46
	Absent	–	35
	Total	3	81
Lieut.-Colonel Sir John St. Clair's Company	Present	2	44
	Absent	2	41
	Total	4	85

The first complete year of the Regiment's *de facto* existence passed quietly away and, in the first days of 1758, the four Lieutenant-Colonels received the compliment of being appointed Colonels in North America.

CHAPTER II

1757–1758

SIEGE AND CAPTURE OF LOUISBOURG—REPULSE AT TICONDEROGA — CAPTURE OF FORT FRONTENAC AND OF FORT DUQUESNE

During the winter Loudoun had laid his plans for the next campaign, which included three distinct operations: first, the siege of Louisbourg; secondly, an advance upon Fort Duquesne; and thirdly, an invasion of Canada by way of Lake Champlain.

The strategy of this plan of advance upon three distinct and entirely disconnected lines is open to criticism. The respective distances in a straight line of Fort Duquesne, Montreal, and Louisbourg from New York were about 300, 330 and 840 miles. And, supposing the objects to have been attained, the British forces would find themselves occupying isolated positions upon exterior lines and holding a front, from left to right, of nearly 1000 miles. Correct strategy would have occupied Fort Duquesne and gained possession of the near shores of Lakes Ontario and Erie. The left flank of the British army would, in that case, have been secure for an advance upon Montreal by Lake Champlain. The capture of Louisbourg might have been treated as a secondary consideration and would probably soon have followed the capture of Montreal. The gulf and mouth of the

St. Lawrence should have been blockaded by the British fleet.

The scheme was, however, fully approved in England, but Loudoun was not fated to carry it out. In the previous July, William Pitt the elder had risen to supreme power in England; and in December he recalled Loudoun, appointing General James Abercromby, the nominee of his predecessor in office, to succeed him, and at the same time to take his place as Colonel-in-Chief of the 60th. The first strategical object, viz. the siege of Louisbourg, was to be entrusted to Colonel Jeffery Amherst, at present of the 15th Regiment, but destined, in a few months' time, to succeed Abercromby as Colonel-in-Chief of the Royal Americans, and to hold that appointment for nearly forty years.

It was Pitt's hope that the siege should be begun in April. A powerful fleet under Admiral Boscawen, with a large convoy of transports, sailed from England on February 19, 1758, but through unfavourable weather did not reach Halifax until May 9. Amherst himself, who sailed with Captain George Rodney[1] on March 16, was equally unfortunate in his passage; being so long delayed that Boscawen, on May 28, weighed anchor, and was sailing to Louisbourg without him, when happily he met Rodney's ship just outside Halifax harbour. The whole fleet, numbering in all one hundred and fifty-seven sail, then steered eastward for its appointed destination.

'The stormy coast of Cape Breton,' observes Francis Parkman, the American historian, 'is indented by a small land-locked bay between which and the ocean lies a tongue of land dotted with a few grazing sheep, and intersected by rows of stone that mark more or less distinctly the lines of what once were streets.' *Fuit*

[1] The future distinguished Admiral. In 1759 he destroyed the stores collected at Havre for the invasion of England. In 1780 he defeated the Spanish fleet off Cape St. Vincent. In 1782 he defeated the French in the West Indies.

Ilium. 'Green mounds and embankments of earth enclose the whole space, and beneath the highest of them yawn arches and caverns of ancient masonry. This grassy solitude was once the Dunkirk of America; the vaulted caverns where the sheep find shelter from the rain were casemates in which terrified women sought refuge from storms of shot and shell, and the shapeless green mounds were citadel, bastion, rampart and glacis. Here stood LOUISBOURG; and not all the efforts of its conquerors, nor all the havoc of succeeding times, have availed to efface it. Men in hundreds toiled for months with lever, spade and gunpowder in the work of destruction, and for more than a century it served as a stone quarry, but the remains of its vast defences still tell their tale of human valour and human woe.

'At the beginning of June 1758, the place wore another aspect. Since the Peace of Aix-la-Chapelle vast sums had been spent in repairing and strengthening it, and Louisbourg was the strongest fortress in French or British America. Nevertheless it had its weaknesses. The original plan of the works had not been fully carried out; and owing, it is said, to the bad quality of the mortar, the masonry of the ramparts was in so poor a condition that it had been replaced in some parts with fascines. The circuit of the fortifications was more than a mile and a half, and the town contained about four thousand inhabitants.'

The brigadiers appointed for Amherst's force were three, namely, Colonels Whitmore, Lawrence of the 60th, and the celebrated James Wolfe.[1] The Right Brigade under Whitmore consisted of the 1st Battalion of the Royal Scots, the 40th, the 3rd Battalion of the 60th under the command of Lieut.-Colonel Young or Major Augustin Prevost, the 48th and the 22nd. The Centre Brigade under Wolfe included the 17th, the 47th, the 2nd Battalion of the 60th under Colonel the Hon. R. Monckton, its Colonel-Commandant,[2] and the 35th. The

[1] Wolfe, 1726–1759. Fought at Dettingen, Fontenoy, Colloden, Maesdricht, Rochfort. Commanded a battalion at the age of twenty-three.

[2] At this period several instances occur of Colonels-Commandant exercising the executive command of their battalions. Sir J. St. Clair, the Lieut.-Colonel, would seem to have been always employed on the staff of the army. Robertson, the Major, was Deputy-Quartermaster-General.

THE HON. ROBERT MONCKTON

(From an engraving after the portrait by Benjamin West, P.R.A.)

Left Brigade under Lawrence was made up of the 28th, 58th, Fraser's Highlanders, the 45th, and the 15th.[1] The force numbered in all some eleven thousand men. The British War Office in those days, and for many years after, ignored the use of Light Infantry regiments or even companies; but as a matter of fact it was always found necessary to improvise light troops for every campaign; and on the present occasion five hundred and fifty marksmen, including seventy or eighty of the 60th, had been drafted together from the different British corps, and placed under the command of a certain Major Scott, with the title of the Light Infantry. They were chosen

'out of the most active, resolute men from all the battalions of Regulars; dressed some in blue, some in green jackets and drawers, for the easier brushing through the woods: with ruffs of black bear skin round their necks; the beard of their upper lips, some grown into whiskers, others not so, but well smutted on that part. . . . Their arms were a Fusil (or Rifle) Cartouche box and a Powder Horn.'[2]

Thus early in its career a part of the 60th discarded the conventional red uniform. We shall presently see that Bouquet, at this very time, was giving much thought to this same question of light troops; but he had not the good fortune to be present, for his services were being utilised elsewhere.

[1] The order of the regiments is given as they stood from right to left. It will be observed that in the right and centre brigades the right has the first place of honour, then the left, then the right centre, then the left centre. Last came the exact centre. In the left brigade the seniority is reversed, the left being the first place of honour, then the right, then the left centre, and then the right.

The following remarkable Brigade Order was issued by Lawrence:

'Experience having discovered that ginger and sugar mixed with the water of America prevent the effects of it, and preserve the men from fevers and fluxes better than anything else yet found out, Br.-Genl. Lawrence does therefore recommend the use of this discovery to the troops in the strongest manner.'

[2] *An Authentic Account of the Reduction of Louisbourg* (London: 1758).

The strength of the two battalions of the Royal Americans was as follows:

2nd Batt. . 39 Officers 984 N.C.O.s and men;
3rd Batt. . 35 Officers 866 N.C.O.s and men.

The harbour of Louisbourg is a bay with an extreme width of about two and a half miles. The entrance, however, is but a mile wide, and is narrowed practically to one half the distance by a chain of rocky islets. All ingress to a fleet was barred by a battery on an island on the western side of the channel, and by a second battery on the eastern shore at Lighthouse Point, while the fortress itself stood on a triangular peninsula on the western shore. Being more vulnerable on the side of the land, the western face, was, of course, the most formidable. It was fortified by four bastions; named, from north to south, the Dauphin's, the King's, the Queen's and the Princess'. The King's bastion formed part of the citadel. Its glacis sloped down to a marsh which completely protected the front, but at both extremities of the line rising ground gave favourable positions for the batteries of a besieging force. The fortress mounted in all two hundred and nineteen pieces of cannon and seventeen mortars; while five ships of the line and seven frigates, fully manned, rode at anchor in the harbour. The garrison under the Chevalier de Drucour, the Governor, consisted of four thousand French regular troops,[1] with several companies of colonials from Canada. The task assigned to Colonel Amherst was certainly formidable.

His first object was, of course, to effect a landing. For this there were three possible places: Freshwater Bay, about three miles west of the town; Flat Point, a little

[1] The second battalions of the following regiments: Volontaires Étrangers, Cambis, Artois, Bourgogne, and twenty-four companies of Marines.

nearer to it; and White Point, which was within a mile of the ramparts. East of the fortress there was yet another place near Lorambec. It was determined to threaten each of these points simultaneously—Lawrence's brigade moving upon White Point, with one regiment detached against Lorambec; Whitmore's brigade menacing Flat Point; while Wolfe delivered the true attack upon Freshwater Cove. No very clear information regarding the French defences erected to cover these landing-places had been received; and the event proved that the point chosen for disembarkation was the most strongly defended of all. Amherst had, however, two advantages; the first that, except off White Point, water was deep enough to permit frigates to ply within range of the land; the second, that he had a sufficient number of boats to carry at least one half of his force.

The General arrived before Louisbourg on June 2, but foul weather for six days prevented him attempting to land. At two o'clock on the morning of the 8th the troops got into the boats; while seven frigates standing in opened a fierce cannonade upon the three threatened points. Wolfe's first division consisted of the five senior companies of grenadiers, the Light Infantry, and a body of American Irregulars, with Fraser's Highlanders and the rest of the grenadier companies, including the two of the Royal Americans, in support.[1] The beach at Freshwater Cove was five hundred yards long with rocks at each end; and on the shore above it over a thousand Frenchmen lay hidden in entrenchments covered by strong abattis. Eight pieces of cannon, great and small, cunningly masked by evergreen shrubs planted in front of them, were in position to sweep every part of the beach

[1] Major Prevost was one of the Field Officers detailed for this attack.

and enfilade the approaches; three of them being so disposed as to rake the cove from end to end.

On receipt of a signal from Wolfe 'the Light Infantry, Highlanders and Grenadiers intermixed, rushed forward with impetuous emulation, . . . and piqued themselves mightily which boat could be most dexterous and active in getting first on shore.'[1] The French remained quiet until the boats were close to the land, when on a sudden they opened a tremendous fire of grape and musketry. Wolfe, thinking success hopeless in the face of such a tempest of shot, signalled to the flotilla to retire; but not observing, or disregarding, the signal three boats of the Light Infantry on the extreme right—one in charge of Lieutenant Brown, another of Ensign Grant, both of the 60th—being little exposed to the fire, pulled in to a craggy point just eastward of the beach, which was sheltered from the enemy's cannon by a small projecting spit. Major Scott's boat was the next to touch land, and was at once stove in against the rocks; but he scrambled out, and with only ten men held his own until supported. Wolfe at once revoked his order and hastened the rest of his force to the same spot. Most of the boats were upset, many men were drowned, but the remainder made their way ashore. After landing they were obliged 'to scramble up such rugged rocks and almost perpendicular precipices as to the wary enemy's engineers seemed in need of no fortification of defence, their own steep, rough ascent having been judged beyond the attempt of men under arms. They had succeeded in gaining what had till now been thought an inaccessible shore, and landing in the most unexpected—one who had not the strongest proofs of the fact might say—incredible places.' Then, forming on the firm ground beyond the

[1] *An Authentic Account of the Reduction of Louisbourg.*

rocks, they swept the French out of the nearest battery by a single rush upon their flank. Lawrence's brigade at once rowed to the same spot, and landed practically unopposed at the western end of the cove. Colonel Amherst followed in time with more troops, and the French, fearing to be cut off from the town, abandoned their guns and fled. Amherst pursued them through what he described as the roughest and worst ground that he had ever seen, until checked by the fire from the ramparts of Louisbourg. The British loss, which fell chiefly on the grenadiers of the 15th, was about 130 of all ranks, killed, wounded and drowned. That of the French was far greater, including 70 prisoners; and having fortified the shore for miles, they left no fewer than 33 guns, great and small, in the hands of the British.

During the two following days the weather was so inclement that, beyond a few tents, little could be landed for the army on shore; but on the 11th the weather moderated and a few light guns were disembarked. Next day (12th) Amherst detached Wolfe with about fifteen hundred men, including 190 of the Royal Americans, to Lighthouse Point, which had been abandoned by the French, in order to fire thence upon the Island battery and the ships in the harbour. Continual high winds prevented for some days the landing of the heavy guns necessary for this post; but on the 20th Wolfe opened fire, and during the 25th the Island battery was silenced and the French men-of-war driven under the fire of the main fortress. Thus the harbour was opened to the British fleet; whereupon the French commander on the night of the 29th contrived under cover of a heavy fire to sink six large ships in the channel, and sealed it up anew.

Meanwhile, Amherst, having secured his own position by field fortifications, had been busy in making a road from the landing-place at Flat Point Cove to a hillock off the north-western angle of the fortress, whence he had determined to deliver his attack. This was a most laborious work, the ground being a sea of mud; and the road moreover required to be defiladed from a frigate, the *Aréthuse,* which was anchored in the Barachois. By the 25th, however, the work was well advanced, and the first trenches were opened towards the Barachois itself. A few days later Lighthouse Point was taken over by the Marines; and Wolfe, with his detachment, broke ground opposite the Princess' bastion.[1] Another ten days saw the first batteries completed, and on July 14th, Wolfe, who had meanwhile been shifted to the north of the town, made a dash at nightfall upon some rising ground opposite to the Dauphin's bastion, and took possession of it. The British cannonade then became heavy and continuous. On the 21st, three of the French line-of-battle ships were set on fire and burned to the water's edge. 'By this time the position of the besieged was deplorable. Nearly a quarter of their number were in the hospitals, while the rest, exhausted by incessant toil, could find no place in which to snatch an hour of sleep.' 'And yet,' says an officer, 'they still show ardour . . .' On the front of the town only four guns could fire at all. The rest were either dismounted or silenced by the musketry from the trenches. The masonry of the ramparts had been shaken by the concussion of their own guns; and now, in the Dauphin's and King's bastions, the English shot brought it down in masses. On the rising ground at the right the trenches had been pushed so close that a great part of the covered

[1] Giving vent to their surprise at the sudden and effective nature of Wolfe's strokes, the garrison used to say, 'There is no certainty where to find him, but wherever he goes he carries with him a mortar in one pocket and a 24-pounder in the other.'

MEDAL AWARDED FOR THE CAPTURE OF LOUISBOURG

way was enfiladed, while a battery on a hill across the harbour swept the whole front with a flanking fire. Amherst had ordered the gunners to spare the houses of the town; but according to French accounts, the order had little effect, for shot and shell fell everywhere. 'There is not a house in the place,' says the diarist just quoted, 'that has not felt the effects of this formidable artillery fire.' Drucour now felt further resistance impossible. On the 26th, the fortress, after a most gallant defence, surrendered. The prisoners taken from the French numbered over five thousand soldiers and seamen, while their casualties in the siege must have been heavy both from sickness and from the British fire. Two hundred and thirty-one guns and seven mortars were found in the place. Amherst's losses were 170 of all ranks killed and 350 wounded. Of the Royal Americans the 2nd Battalion had Lieutenant Hay and 7 men killed and 14 men wounded; and the 3rd Battalion, 17 men killed and 43 wounded.

The result of this success was the capture of the Island of Cape Breton; and Amherst,[1] urged on by Wolfe, then proposed to Boscawen to sail to the attack of Quebec; but the Admiral judged the enterprise to be impracticable; so the force was broken up into three columns to complete the conquest of Prince Edward's Island, the Bay of Fundy, and the Gulf of St. Lawrence. Of these columns, one, under Colonel Monckton, comprised his own battalion, with the 35th, and was employed in the destruction of the French settlements in the River St. John at the head of the Bay of Fundy. Monckton had not yet finished his task by November, when he sailed for Halifax, whither our 3rd Battalion had already preceded him.

[1] 'Mr. Amherst has displayed the General in all his proceedings,' writes Capt. Knox of the 47th in his Journal.

We must now follow the fortunes of General Abercromby's force, designed during this summer for the attack on Ticonderoga—the third of the objectives mentioned on page 44. For this enterprise had been allotted between six and seven thousand regular, and nine thousand provincial, soldiers; the former consisting of the 27th, 42nd, 54th, 46th, 55th, the 4th Battalion of the Royal Americans, six companies of the 1st Battalion,[1] and a newly raised corps known as Gage's Light Infantry. The early months of the summer were consumed in the laborious task of transporting supplies to the head of Lake George, where the entire force was finally assembled at the end of June. On July 4 the stores were shipped, and on the following day the troops embarked in the flotilla that was to convey them across the lake. There were in all over a thousand boats of one kind and another, which sailed down-stream under a cloudless sky in three parallel columns; that in the centre conveying the British regular battalions. At the head of this central column was Lord Howe, late Colonel-Commandant of our 3rd Battalion, but now Colonel of the 55th—an officer so progressive in his ideas that he had served with the backwoodsmen to learn forest warfare, and had insisted on cutting short the hair of his men, shearing the skirts off their coats, browning the barrels of their muskets, and generally equipping them in a sensible and practical fashion. Though Abercromby—described as 'a heavy man'—was nominally the commander-in-chief, Howe, termed by Wolfe 'the best soldier in the British Army,' was the true leader alike in intellect and energy.

The vanguard of the naval column was found by the 55th and Royal Americans. By daybreak of the 6th

[1] First Battalion 60th, 562 rank and file; say 630 of all ranks. Fourth Battalion 60th, 932 rank and file; say 1050 of all ranks.

the flotilla had passed the lake, and reached the narrow channel that leads into Lake Champlain by the headland of Ticonderoga.

'The spectacle,' says Parkman,[1] 'was superb: the brightness of the summer day; the romantic beauty of the scenery; the sheen and the sparkle of those crystal waters; the countless islets, tufted with pine, birch and fir; the bordering mountains with their green summits and sunny crags; the flash of oars and glitter of weapons; the banners, the varied uniforms, and the notes of bugle, trumpet, bagpipe and drum, answered and prolonged by a hundred woodland echoes.'

A French advanced party on the shore was driven back, and the work of disembarkation went rapidly forward so that by noon the entire army had been landed on the western shore of the lake. Some Colonial Rangers were sent forward to reconnoitre; and the troops, formed in four columns, plunged into the virgin forest for the advance; the idea being to follow the western bank of the channel which joins Lake George to Lake Champlain. But the undergrowth being thick and the ground choked by the decaying trunks and limbs of fallen trees, orderly array proved impossible; the columns became intermixed, guides lost their track and their heads; and the army was for a time hopelessly astray in the forest. Meanwhile the French party, fearing to be cut off, had also plunged into the woods and lost its bearings; and came into sudden collision with Howe and his Rangers. The French fired a volley which killed Lord Howe. The Rangers although outnumbered by nearly two to one stood firm, but at the report of mus'etry the regiments behind fell back. They were soon rallied, and the enemy, attacked in rear by other scouts, was almost destroyed, 140 being killed and 152 captured out of 350 men. But

[1] *Montcalm and Wolfe*, ii. 96.

with the death of Lord Howe the guiding spirit of the British force had disappeared.

Hopelessly bewildered, Abercromby halted as many of his men as he could collect; and keeping them under arms for the night, fell back next day to his landing-place, where he was relieved to find the rest of his force awaiting him. Meanwhile the Marquis de Montcalm, who was in command of the French, had realised the peril of his own situation. The channel between the two lakes being impassable owing to rapids, the usual route to Ticonderoga ran over a bridge by some saw-mills at the foot of these rapids. This bridge Montcalm had destroyed, having determined to risk a fight on the western bank of the channel; and it was not until the evening of the 6th that, upon the advice of two of his officers, he fell back to Ticonderoga itself. On the 7th Abercromby sent Lieut.-Colonel Bradstreet[1] of the 60th with the 1st Battalion to occupy the saw-mills; and, the bridge having been rebuilt, the army encamped on the ground lately occupied by the French, within two miles of the fort.

The peninsula of Ticonderoga consists of a rocky plateau with low ground on each side, standing at the junction of the two lakes. The fort lay at its eastern extremity; but half a mile to westward of it the ground rises and forms a ridge across the plateau; and upon this ridge Montcalm, after considerable hesitation, decided at the last moment to accept battle. By dawn of the 7th every man of his force was at work felling the trees before it. The logs were piled into a massive breast-work, nine feet high and loopholed; the trees in advance were also cut down, and left lying with their heads outward; and between them and the breast-work itself the ground was covered by heavy branches, with sharpened points, and

[1] Bradstreet seems to have acted as Quartermaster-General.

boughs interlaced into a formidable abattis. But in spite of all this the position was not really formidable. There was a hill called Mount Defiance upon its southern flank from which cannon could have raked it from end to end; while if Abercromby had bethought him to mask this position with a part of his force, he could have pushed the rest past it to Lake Champlain, and cut off Montcalm's supplies and retreat beyond all succour. Against Abercromby's 15,000 men, the French commander had but 3600,[1] and provisions for only eight days. Opposed by a leader of ability his position would have been hopeless.

But Abercromby, unlike his more famous namesake Sir Ralph, was not a genius. He directed his chief engineer, Lieutenant Clarke—a mere boy—to reconnoitre the French position; but that officer and his assistants did their work very perfunctorily and returned with conflicting reports. Clarke himself treated Montcalm's fortifications as formidable only in appearance; and although Abercromby had left his guns at the landing place and believed himself opposed by 6000 men, yet influenced by the information given by his prisoners to the effect that the enemy was hourly expecting a reinforcement of 3000, he neglected the duty of examining the works with his own eyes, and determined to storm them off-hand with the bayonet.

The French army was posted as follows. On the left under General Bourlamaque were the battalions of La Sarre and Languedoc. The centre under Montcalm in person comprised a battalion of Berry and one of Royal Roussillon; those of La Reine, Béarn, and Guienne under the Chevalier de Lévis, second-in-command, formed

[1] These are the numbers given by Montcalm. The Marquis de Vandreuil, influenced probably by jealousy, puts Montcalm's force at 5000.

the right. The grenadiers and the 2nd Battalion of Berry were held in reserve. The low grounds between the breast-work and the outlet of Lake George were occupied by volunteers; while the foot of the hill on the side of Lake Champlain was held by 450 men of the Colonial regular troops.

On the 8th skirmishing between the irregulars on either side began at 9 A.M., and continued without interrupting Montcalm's occupation of strengthening his position. It was not till noon that volleys of musketry from the forest in front announced that the British light troops, under Bradstreet, were driving in the French outposts. At about 1 P.M. columns of attack were observed; the picquets leading the way with the massed grenadiers of the entire force in support—the whole under Colonel Haldimand, who had exchanged with Sir J. St. Clair and was now Commanding Officer of the 4th Battalion. The regiments of the main body followed the grenadiers, excepting the 42nd and the 55th which were held in reserve. Gleaming in the sunlight the columns of red uniforms advanced to the assault of the parapets, the tops of which only those in front could discern, and behind whose shelter the French garrison lay invisible. Nearer and nearer they approached till, on a sudden, a withering volley struck the leading files to the ground. Still the columns advanced, and the abattis guarding the entrenchments were reached; but the mass of pointed branches first checked their pace, then broke their ranks, and all the while the struggling mass was raked on either flank by deadly cross fires. After the action Colonel Haldimand wrote in a letter that he had never seen so hot a fire borne with such unflinching courage. But it was more than flesh and blood could stand. Declaring that the task was impossible the gallant troops at length fell back.

The general, watching the scene from a certain distance, ordered a renewal of the attack; and the decimated but undaunted columns once more advanced with a cheer. But the result was the same.

> 'The scene was frightful. Masses of infuriated men who could not go forward and would not go back, straining for an enemy they could not reach, and firing on an enemy they could not see, caught in the entanglement of fallen trees; tripped by briars, stumbling over logs; tearing through boughs; shouting, yelling, cursing; and pelted all the while with bullets that killed them by scores, stretched them on the ground, or hung them on jagged branches in strange attitudes of death. The provincials [1] supported the regulars with spirit and some of them found their way to the foot of the wooden wall.' [2]

An attempt by Colonel Bradstreet to turn the enemy's left by boats full of troops on the river was repulsed, and even the boats were sunk by the enemy's guns. Behind their breast-works the French in their white uniforms kept up an increasing fire and fought with cheerful confidence. At every point where his presence was required Montcalm animated his troops, albeit—as his despatches show—lost in admiration of the British columns who, between 1 and 6 P.M. attacked him no less than six successive times. Once a party of Highlanders actually surmounted the parapet, only to fall on the bayonets of the defenders. Once a handkerchief appeared above the earth-work and the assailants hailed an unexpected surrender. With shouts of 'Quarter' they advanced to the entrenchments. The thoughtless signaller had, however, meant to indicate, not surrender but defiance; and the French, observing the approach of the shattered columns with every sign of cessation of strife, believed

[1] The York Regiment seems to have been the only one to carry its support to the extent of self-exposure. The others fired into the British troops from behind.

[2] Parkman.

on their part that the British desired to capitulate, and ceased their fire. A French officer quickly rectified the error. Fire was recommenced, and a volley struck some of the British to the ground and drove back the remainder.[1]

The attacks continued till dusk, the 4th Battalion of the 60th being conspicuous for its steadiness and pertinacity.[2] At length the General resolved to abandon the contest. But it was not until repeated orders had been given that the Royal Americans could be induced to leave the ground. Even then skirmishing continued until dark to cover the retirement and carry off the wounded. Abercromby gave orders to retreat to the landing-place. But after the terrible strain to which the troops had been exposed, the reaction was too great, and confusion at length ensued. Had it not been for the resolution of Colonel Bradstreet, who set a guard over the boats and forced the fugitives to orderly embarkation, numbers would have perished by drowning.

Abercromby hastened back with all possible speed to the head of Lake George, and there entrenched himself —the most thoroughly demoralised man in the whole army.

The losses of the British in this disastrous affair were very heavy, amounting to nearly 2000 men, of whom over 1600 belonged to the seven battalions of regular troops. In the 1st Battalion of the 60th Captain-Lieutenant Forbes, Lieutenant Davis and 21 men were killed; Major Tullikens, Captains Baron Dietrich

[1] The incident recalls similar ones in the South African war. On the present occasion the British were loud in their complaints of French treachery; but it does not appear that any was intended. Some say the handkerchief was red not white.

[2] This is Patrick Murray's statement, who adds that the Battalion was commanded by Baron Münster. This officer is however, returned among the wounded of the 1st Battalion, whose losses were proportionately greater than these of the 4th.

Herbert Münster, Mather, and Cochrane, Lieutenants Barnsley, Ridge, Wilson and Guy; Ensigns Baillie, Gordon and McIntosh, and 86 men were wounded. In the 4th Battalion Major Rutherford, Lieutenant Hazlewood and 25 men were killed; Lieutenant-Colonel Haldimand, Captains Prevost and Du Fez, Captain-Lieutenant Schlösser, Lieutenants McLean, Allaz, Turnbull, and McIntosh, and 126 men were wounded. Altogether it was a murderous fight; and men who had fought at Fontenoy declared that the fire there was child's play compared with that at Ticonderoga. The loss of the French in the battle amounted to three hundred and forty-seven killed and wounded: among the latter being Generals Bourlamaque and Bougainville.

The repulse effectually checked all advance by way of Lake Champlain, and the operations in that quarter were suspended for several weeks. Indeed the initiative lay with Montcalm. Abercromby was demoralised and, but for the enormous disparity of numbers, British territory would have been in danger of French invasion. Just at this critical moment, however, the resourceful Bradstreet turned the scale. While others almost despaired of successful defence, his mind was turned to resuming the offensive. He warmly advocated, and by dint of untiring pressure obtained the consent of General Abercromby to, an enterprise of which Lord Loudoun had already approved: viz. an expedition against Fort Frontenac (the modern Kingston), an important post situated on the north bank of the strait connecting Lake Ontario with the St. Lawrence. It commanded the lake and had formed a starting-point for Montcalm in his attack on Oswego. On August 22 Bradstreet set out in his flotilla of boats on Lake Ontario, with three thousand men, all provincials except two companies of the 60th (comprising 8 officers and 146 men). Three days later he landed near the fort. Its commandant was M. Payau de Noyau, a poet, but none the less a gallant veteran. De Noyau had foreseen the danger, and having only 110 men had asked Vaudreuil for a reinforcement. The reinforcement had arrived. It consisted of a single man—a fraction of one, to be quite accurate, for the reinforcement had only one arm. During the night of the 26th Bradstreet effected a lodgment within two hundred yards of the fort, and de Noyau, finding his post untenable, surrendered. With the garrison was taken the whole French naval force on Lake Ontario, consisting of nine vessels each carrying from eight to eighteen guns. The crews unfortunately escaped ; but an enormous quantity of munitions of war and stores fell into our hands : and in the fort itself were found sixty pieces of cannon and sixteen mortars. The blow struck was second only to the capture of Louisbourg, and had Bradstreet been supported by a force to re-occupy Oswego and thus recover a harbour on Lake Ontario, the captured vessels could have been taken there and the consequences of the victory would have been even greater than they actually were. But, even as it was, New France was cut in two : Fort Duquesne was left isolated, and the Five Nations once more turned towards England. It was the practice of the French when they took prisoners to turn away and allow their Indians to scalp the captives. The Oneidas pressed for the same privilege on the present occasion, but, happily for the credit of the Regiment, Bradstreet sternly refused.

Thus ended Abercromby's campaign ; for, although reinforced early in October by five battalions under Amherst in person, the two generals decided that the season was too much advanced for a renewed attack upon

DIETRICH HERBERT FREIHERR VON MÜNSTER

(From a family portrait presented to the King's Royal Rifle Corps by Prince Alexander Münster, 1912)

(The picture from which this reproduction was made was painted at the age of 18.)

The Freiherr von Münster was born in 1728; joined the British service as Captain in the Royal American Regiment, 1756; afterwards Major-General and Governor of Minorca. Died 1776.

Ticonderoga. Drucour had lost Louisbourg, but his obstinate defence had for the moment saved Canada.

The second objective mentioned on page 44 was the capture of the redoubtable Fort Duquesne. The force employed for the purpose was placed under command of Brigadier-General Forbes. It was about 6000 strong, of whom three-fourths were provincials, and included the four companies of our 1st Battalion (363 strong) under Colonel Bouquet, which had not been at Ticonderoga, and whose task had been that of defending the western frontier of British territory. Forbes was an old soldier who had seen much service, and had thought out for himself the true method of carrying on a forest campaign. Warned by Braddock's failure, he resolved not to encumber himself by an unwieldy train of wagons, but to establish fortified depots at intervals of forty miles, and, when arrived with striking distance of his point of attack, to march with all his force and with as few encumbrances as possible. He had original ideas also as to the training and equipment of troops, which he duly communicated to Colonel Bouquet. 'We must learn the art of war from the Indians,' he wrote; and Bouquet was so thoroughly of his mind that, like George Washington, he wished to dress his men like the Indians. But Bouquet did not stop at mere matters of dress, for he obtained sixteen rifled carbines or, as they were called, rifle-barrelled 'fuzils' or 'fusees,' for his battalion, and thus turned them into riflemen before their time. Moreover he invented a new exercise for making his troops handy at forest fighting, sending them into the woods in small columns with two men abreast, 'which deployed into line in two minutes with the light troops *en écharpe*.'[1] It is

[1] *Vide* Appendix IV.

reasonable therefore to assume that the companies of his battalion attached to the expeditionary force were exceptionally well trained.

With the exception of a strong battalion of Montgomery's Highlanders the rest of the troops allotted to Forbes were provincial, and therefore not easily forthcoming at short notice. The base of operations was Philadelphia, where Forbes in person arrived during the month of April; but the end of June had come before the contingents which composed his army were ready to march. Then came a wrangle over the route which should be chosen. Virginia was furiously jealous lest a trading road should be opened between Pennsylvania and the Ohio; and young George Washington, being a Virginian, was most urgent that the other track, namely that taken by Braddock, should be followed. But Forbes cared nothing for provincial squabbles and, at the instance of Bouquet and Sir J. St. Clair, Q.M.G., another Lieutenant-Colonel of the 60th, decided upon cutting a new road, and making a shorter line of advance from Pennsylvania. Even so the whole distance from Philadelphia to Fort Duquesne would be 324 miles. Washington was furious; but the event showed that, in spite of his familiarity with the country, he was inferior to Bouquet in judgment and military experience.

At the beginning of July Colonel Bouquet, with an advanced party, encamped at Raystown, on the eastern slope of the Alleghanies, which was the first of the fortified posts. The work being named Fort Bedford, the town retains the name of Bedford to this day. At the same time Forbes moved up to the frontier village of Carlisle, and thence to Shippensburg, where an illness from which he was already suffering increased with pain so excruciating that he was unable to advance farther until September. The distance from Philadelphia to Carlisle was 121 miles; from Carlisle to Raystown, 98. At Raystown the advance party was 105 miles from Fort Duquesne. By Braddock's road the distance would have been 145 miles.

In the meanwhile Bouquet, with immense labour and in the face of the greatest difficulties—for forage was scarce, defiles and flooded rivers abounded—was pushing forward the construction of a road towards the next fort at Loyalhannon Creek, 49 miles distant. This necessarily took much time, but the delay was not wholly without advantage, for the French had already summoned the friendly Indians to Fort Duquesne, and if the attack upon it were postponed, those fickle allies would most probably grow tired of waiting, and would disperse to their homes. Forbes had already sent a confidential emissary to warn them that the hour of France was on the wane, and that their true interests lay in alliance with the English.

> 'Meanwhile,' says Parkman, 'Bouquet's men pushed on the heavy work of road-making up the main range of the Alleghanies; and, what proved far worse, the parallel mountain ridge of Laurel Hill; hewing, digging, blasting, laying fascines and gabions to support the track along the sides of steep declivities, or worming their way like moles through the jungle of swamp and forest. Forbes described the country to Pitt as an "immense, uninhabited wilderness, overgrown with trees and brushwood, so that no one can see twenty yards." In truth, as far as eye or mind could reach, a prodigious forest vegetation spread its impervious canopy over hill, valley and plain, and wrapped the stern and awful waste in the shadows of the tomb.' [1]

But patience and determination surmounted all difficulties. Forbes, in a letter of August 28, after com-

[1] *Montcalm and Wolfe*, ii. 147.

plaining of the 'rascality of the country people' in the non-fulfilment of their contracts, writes:

'Everything that depended upon the troops has succeeded to admiration and we have got entirely the better of that impossible road over the Alleghany mountains and Laurell ridge. A few petty skirmishes has resulted in the taking and losing of a few scalps.'

General Forbes promised to send his correspondent a couple as a little memento of the expedition. With the object of gaining intelligence the service of a few Indians was secured: but their insolence overcame the patience of all but Bouquet, to whom Sir J. St. Clair, the Q.M.G., whose temper was not of the best, wrote: 'The greatest curse that our Lord can pronounce against the worst of sinners is to give them business to do with provincial commissioners [1] and friendly Indians.'

The advance guard and the task of clearing a road through the forest was assigned by Bouquet to Colonel J. Burd of the Pennsylvania Regiment. Burd seems to have been a good man and may have kept his battalion in some sort of order; but as a rule the provincial troops were unruly, undisciplined, and almost unarmed. In August Bouquet directed Burd to advance to Loyalhannon (Fort Ligonier) with the Royal Americans, the Highland Battalion, a division of Artillery, and his own battalion. Three days' provisions and entrenching tools were to be carried, and wagons of further provisions taken as far as the road had been constructed. The rest of Bouquet's letter of instructions deserves to be copied verbatim.

'The Horses are to be tyed every night upon the mountain, as they would otherwise be lost; *Locus* is to be cut for them. They could perhaps be left loose at Edmund Swamp and Kickeny Pawlins.

Lieut. Chew with a Party are to be detached from the Top of

[1] The provincial authorities annoyed even Bouquet, who writes: 'Je n'y ai rien appris de satisfaisant. La plupart de ces Monsieurs ne connoissent pas la différence d'une party et d'une armée, et trouvent facile tout ce qui flatte leurs idées, sautant par dessus toutes les difficultés.'

the Allegheny to reconnoitre in a straight Line the ground betwixt that place and the Gap of Lawrell Hill, he is to cross that Gap, observing the course of the Water, and the Path; and is to join the detachment at L.H. All the detachments of the R.A.R. those of the 5 Comps of Highlanders, and of your own Battn are to march with you to Loyal H. Col. Stephens is to march with you and his six compys with 3 or 4 days Provisions for the whole. At the place where you leave the Artillery and Waggons your men are to carry the Tools themselves, Packing on the Horses the Saws Grindstones &c.

You are to employ all the Pack Horses of the first Batt. and those that you may find on the Road to carry your Provisions until the Waggons can come to you, and load the 5 Barrels of Cartridges; Drive also some Bullocks.

As soon as you arrive at L.H. Mr. Basser is to lay out your Incampment at the place assigned by Mr. Rhor, with two small Redoubts at 200 yards; All hands are then to be employed entrenching the camp; Those who have no Tools will pitch the Tents, cook and the rest releave one another in the work.

Before night the Ground must be reconnoitred and your advanced guards posted; The Centrys are to relieve every hour in the night without noise.

No Drum is to beat as long as you judge that the Post has not been reconnoitred by the Enemy.

Suffer (in the beginning chiefly) no hunters or stragglers, to prevent their being taken—No gun to be fired.

A Storehouse of 120 feet long and at least 25 wide is to be built immediately to lodge your Provisions and Ammunition, in the place where the Fort is to be erected, and covered with Shingles.

All the artificers are to be put to work; the Sayiors and Shingle Makers with the Smiths first: an hospital is to be built near the Fort, and ovens, Mr. Rhor is to give the direction for the Fort.

If there is any possibility of making Hay, no time is to be lost, and the clear grounds are to be kept for that use, and not serve for Pasture. Send proper People to reconnoitre where sea Coal could be got, if there is none, Charcoal must be made.

The houses of office to be kept clean and covered every day.

The ammunition and arms carefully inspected, the arms loaded with a running Ball.

The Tools to be delivered to each Party on Receipt of their Commanding officer, who is to see them returned to the Stores before night.

The Intrenchmt is to be divided by tasks, and all the officers are to inspect the Work.

If you send any Party forward, Don't permit them to take scalps, which serves only to render the Ennemys more vigilant. No Party is to be sent untill you hear from Major Armstrong and Capt. Shelby.

It would perhaps be proper to change every day the place of your advanced Posts: secure all avonies.

If any difficulty should occur to you, Consult Major Grant, whose Experience and perfect knowledge of the service, you may rely intirely upon.

I give you the above instructions by way of memorandum and you are at Liberty to make any alterations that your Judgment and the Circumstances may direct.

Let me hear from you every two days. You know that some of the Provincial officers are not vigilant upon guard, Warn them every day. They could ruin all our affairs, Keep a Journal of your Proceedings.

I am Sir Your most obedt. hble. servt.

HENRY BOUQUET.'

Burd advanced in three parallel columns; the Highlanders in the centre being guarded by the Royal Americans on the right and the Provincials on the left. At the end of August the expedition was cheered by the news of the fall of Louisbourg, but in September the rashness of one of Bouquet's officers nearly wrecked the whole expedition. Major Grant of the Highlanders begged for a small force to reconnoitre Fort Duquesne 56 miles distant, and strike some blow which would discomfort and discourage the French. Bouquet, as shown in his letter, had confidence in Grant and made over to him 800 men, including Highlanders, about 100 of the 60th, and Provincials; whereupon Grant left a fourth of his men, including the 60th, as baggage-guard, and with culpable rashness scattered the rest in all directions. One party was sent out to right, another to left, while he himself with 100 men took post in front of the baggage, and sent a company of Highlanders still farther forward to open ground near the fort. The French and Indians at once sallied out and drove all these parties back, one after another. After a gallant resistance the Highlanders broke. The panic soon spread, and but for the firmness of the baggage-guard and the arrival of Bouquet with the Royal Americans, the whole of Grant's force would have been cut to pieces. As it was 273 men were killed, wounded or taken, Grant himself being among the prisoners. The loss of the 60th on this occasion is not recorded, but it was probably small. It included, unfortunately, Ensign Charles Rohr who, as D.A.Q.M.G., had been doing excellent work in planning and superintending the construction of the road over the mountains. Bouquet had evidently a high opinion of him. The Highlanders, who were farthest in advance and inexperienced in bush fighting, were the chief sufferers.

This reverse was discouraging, and the adverse weather added to Forbes's difficulties. Heavy rain destroyed Bouquet's new road; the horses, being underfed and overworked, succumbed in numbers; and the magazines at Bedford and Loyalhannon were emptied faster than they could be replenished. On October 12, the enemy attacked the advanced post at Loyalhannon, but, after a vigorous action, were repulsed with loss. All through October the rain continued, till it gave place to snow.

'Dejected Nature wept and would not be comforted. Above, below, around, all was trickling, oozing, pattering, gushing. In the miserable encampments the starved horses stood steaming in the rain, and the men crouched, disgusted, under their dripping tents, while the drenched picket-guard in the neighbouring forest paced dolefully through black mire and spongy mosses. The rain turned to snow; the descending flakes clung to the many-coloured foliage, or melted from sight in the trench of half-liquid clay that was called a road. The wheels of the waggons sank in it to the hub and to advance or retreat was alike impossible.'[1]

At the beginning of November Forbes, a dying man, was carried to Loyalhannon, where he decided that nothing further could be done until the next spring. But the capture of Fort Frontenac, on August 27, deprived Fort Duquesne of its supplies, and had already decided its fate. Intelligence was brought to Forbes that the French garrison had been so much weakened as to be almost defenceless, and on November 18 he set out in a litter at the head of 2500 picked men—including the Royal Americans, who formed the right flank guard—and a few light guns, without tents or baggage. In the night of the 24th an explosion was heard. The French had blown up the fort and retired to Venango. When the troops reached Duquesne next day they found it a heap of ruins. Forbes, therefore, constructed a stockade round a cluster of huts that was left standing, and named it Pittsburg in honour of the minister.

'If his achievement was not brilliant,' says Parkman, 'its solid value was above praise. It opened the Great West to English enterprise, took from France half her savage allies, and relieved the western borders from the scourge of Indian war. From southern New York to North Carolina, the frontier populations had cause to bless the memory of the stedfast and all-enduring soldier.'

A garrison of 200 provincial soldiers was then left to hold the place; and, after burying the bones of the men who had fallen with Braddock, the force, destitute of tents, shoes and clothing, and living from hand to mouth, marched back to Pennsylvania with the dying Forbes in its midst. He lingered on at Philadelphia until the following March, when he died; but not before he had

[1] Parkman.

caused a gold medal to be struck to commemorate his march, and had authorised his officers to wear it hung round their necks by a blue ribbon.

'So ended the campaign of 1758. The centre of the French had held its own triumphantly at Ticonderoga; but their left had been forced back by the capture of Louisbourg, and their right by that of Fort Duquesne, while their entire right wing had been well nigh cut off by the destruction of Fort Frontenac.'

The internal condition of Canada had become pitiable. The St. Laurence was partly closed by British ships. The harvest had failed. A barrel of flour cost £8. The greater part of the cattle had been killed, and the wretched inhabitants were reduced to a pittance of salt cod. But their loyalty to France shone as brightly as ever. Their chief regret was that arms and ammunition were also failing. It was evident that in default of reinforcement, the end could not be far distant.

The right wing of the 1st Battalion under Bouquet went into winter quarters at Philadelphia; the left wing under Major Tullikens at Albany. The 2nd and 3rd Battalions remained at Louisbourg; the 4th under Haldimand at Forts William Henry and Edward, where it suffered much from cold.

The following memorandum by General Amherst shows the regimental state on January 24, 1759:

Batt.			Wanting to complete
1st	(Brigadier-General Stanwix) (6 Companies)	498	126
2nd	(,, ,, Monckton)	817	223
3rd	(,, ,, Lawrence)	855	185
4th	(Colonel J. Prévost)	797	248

No return of the other four companies of the First battalion. I hear they want 150. JEFF AMHERST.

CHAPTER III

GENERAL WOLFE ATTACKS QUEBEC—REPULSE OF ATTACK ON THE MONTMORENCI HEIGHTS—BATTLE OF THE PLAIN OF ABRAHAM—DEATH OF WOLFE AND OF MONTCALM—OPERATIONS OF AMHERST, PRIDEAUX, AND STANWIX

WHILE Forbes was struggling back from Pittsburg to Philadelphia, the great minister, whose name he had unconsciously immortalised in America, was busy with plans for the next campaign. Abercromby had been recalled. General Amherst was now Commander-in-Chief and consequently Colonel-in-Chief of the Royal Americans.[1] The operations designed for the year 1759 were:

1. A direct attack upon Quebec, for which purpose Amherst was directed to make ten battalions over to Wolfe.

2. An advance under Amherst himself upon Montreal by way of Ticonderoga and Crown Point.

3. Any other enterprise that the General might think fit to undertake without prejudice to the other two.

No fresh troops were to be sent out from England; though some were under orders to sail for Canada from the West Indies; and it may be conjectured that the recruiting parties of the Royal Americans were unusually active

[1] Although Amherst was Colonel-in-Chief, he was junior in the service to two of the Colonels Commandant.

during the winter in order to make good the losses of the past campaign.

The British fleet of twenty-one ships, designed for American waters, sailed from England in the middle of February, under Admirals Saunders, Durell, and Holmes, with Wolfe on board as a passenger. The voyage was long and tedious, but when Louisbourg was reached the harbour was found to be blocked with ice, and Saunders was obliged to make for Halifax. There, it will be remembered, the 2nd and 3rd Battalions of the 60th had been quartered for the winter; and the first business of the fleet was to convoy them and the other regiments of Wolfe's force to Louisbourg, as soon as the ice should permit. By May the concentration was completed; when Wolfe discovered that, instead of twelve thousand men, as he had hoped, he had fewer than nine thousand, though deficiency in numbers was in part made good by excellence in quality. The grenadier companies were as usual massed together, and organised in two battalions. A corps of Light Infantry was also made up of the best marksmen and most promising men from the various regiments, and six companies of provincial rangers provided a useful corps of irregular troops. The remainder were brigaded as follows:

Brigadier Monckton's brigade[1]—15th, 53rd, 58th, and Fraser's Highlanders.

Brigadier Townshend's brigade—28th, 47th, 2nd Battalion of the 60th.

Brigadier Murray's brigade—35th, 48th, 3rd Battalion of the 60th.

The approaching campaign was awaited with anxiety both by the boastful Vaudreuil and the quietly gallant

[1] Monckton was a Colonel-Commandant of the 60th, and Murray became one in the following October.

Montcalm. That Montreal would be the point of attack neither doubted: and consternation reigned supreme when M. Bougainville arrived from France with the news that a great fleet was on its way to attack Quebec. No one had believed that the British would risk the dangerous navigation of the St. Lawrence; and but for Lieutenant Des Barres of the Royal Americans, a talented Engineer who had constructed a chart of the river after the capture of Louisbourg, the attempt might perhaps never have been made.

At the beginning of June the troops embarked, and on the 6th the last division of transports sailed out of Louisbourg for the St. Lawrence. The 2nd Battalion of the 60th, whose Lieutenant-Colonel—Sir J. St. Clair—was Q.M.G. to Amherst, consisted of 27 Officers, 34 N.C.O.s and 520 rank and file under Captain Oswald: while the 3rd Battalion, commanded by Lieut.-Colonel Young, comprised 29 Officers, 34 N.C.O.s, and 544 rank and file.[1] Reinforcements were hurriedly summoned from every part of New France; and after much debate, Montcalm, who commanded in person at Quebec, resolved upon his plans of defence.

Quebec, with its fortifications, stands on the north bank of the St. Lawrence, being situated on a rocky headland which marks the contraction of the river from a width of fifteen or twenty miles to a strait scarcely exceeding one. Immediately northward of this ridge the River St. Charles flows down to the St. Lawrence; and seven miles eastward of the St. Charles the shore is cut by a rocky gorge through which pours the cataract of the Montmorenci. It was between these two streams that Montcalm disposed his army, his right resting on the St. Charles; his left on the Montmorenci; with his

[1] Each battalion had but seven companies present, the other three having been left in Nova Scotia.

headquarters on the little river of Beauport, midway between the two, and his front to the St. Lawrence. All along the border of the great river were thrown up entrenchments, batteries, and redoubts. From Montmorenci to Beauport abrupt and rocky heights raised these defences too high above the water to be reached by the cannon of ships. From Beauport to the St. Charles stretched broad flats of mud which were commanded by batteries both afloat and ashore, as well as by the guns of Quebec. On the walls of the city itself were mounted over one hundred pieces of cannon; a bridge of boats with a strong bridge-head on the eastern side preserved the communication between city and camp; and for the defence of the river itself there was a floating battery of twelve heavy guns, besides several gun-boats and fire-ships. The vessels of the convoy that Durell had failed to intercept, together with the frigates which escorted them, were sent up the river beyond Quebec to be out of harm's way; and the sailors were taken to man the batteries ashore. Thus strongly entrenched, with five regular regiments—La Sarre, Béarn, Guienne, Languedoc, and Royal Roussillon, and a total force—including Militia, Canadians, and Indians—of about seventeen thousand men, Montcalm awaited the approach of Wolfe and his army.

Meanwhile the British fleet had arrived at the mouth of the St. Lawrence, and had spent several days in groping its way up the river. Thanks not a little to the instructions of Des Barres and to the skill and care of James Cook,[1] destined in later life to gain imperishable renown for his voyage round the world, transports and men-of-war alike accomplished the whole passage without a single mishap; and on June 26 the whole armament was anchored off the

[1] James Cook, 1728–1779; discovered Australia, 1770; Sandwich Islands, 1776. Killed by the natives at Owhyhee in the Pacific.

southern shore of the Isle of Orleans, a few miles below Quebec. The troops landed at once without resistance; but on the morrow the Governor of Quebec sent a flotilla of fire-ships down the river upon the British fleet. Thereupon the bluejackets calmly rowed out, grappled the burning vessels, and towed them ashore; while the troops, formed up in order of battle, stood and enjoyed the spectacle. This was the first effort made by the French navy in the St. Lawrence, and it was entirely unsuccessful.

Wolfe, on reconnoitring the French lines, could see no vulnerable point therein; and resolved to begin operations by occupying the heights of Point Lévis,[1] nearly opposite Quebec on the southern shore, and the nearest accessible point whence to open fire on the city. Monckton's brigade having accordingly entrenched itself on these heights threw up batteries, which on July 12 opened fire and greatly damaged the buildings of the town. Meanwhile, on the 8th, Wolfe had sent the brigades of Townshend and Murray to the north bank of the St. Lawrence. They landed at the village of L'Ange Gardien, and after a sharp skirmish entrenched themselves in the forest on the eastern side of the Montmorenci, in the hope of finding some weak spot, or at least of tempting Montcalm to make some movement. But Montcalm, notwithstanding the opportunity that Wolfe offered him by this dangerous division of his force, simply stood on his guard and declined to move. Wolfe searched in vain for a vulnerable point; and the operations came practically to a deadlock, in spite of the fact that the ships on the river exchanged ever incessant cannonade with the batteries on shore, and that constant raids on the British camp were made by Indians and Canadians disguised as Indians. But the raids were repelled, and

[1] Sometimes called Pointe de Lévy.

the Indians complained that the British were learning how to fight. The training of Howe and Bouquet was bearing fruit.

At this juncture, the fleet came forward to show the way. The French had flattered themselves that no vessel could pass the batteries of Quebec; but on July 18 the *Sutherland* of fifty guns, with several smaller vessels, made its way up the river under the fire of Monckton's batteries on Point Lévis, and anchored above the town. Wolfe thus threatened to take Quebec in reverse; but Montcalm only detached some six hundred men to hold the accessible points between the city and Cap Rouge, some eight miles above it, and sat still as before. On the 21st Wolfe despatched the 3rd Battalion of the Royal Americans with the grenadier companies of both battalions (as well as those of the 15th and 48th), which landed twelve miles above the town of Quebec,[1] took about sixty prisoners, and killed several Indians. During the skirmish Major Augustin Prevost was dangerously wounded in the head, but was successfully trepanned.

But Montcalm was not to be moved. A second attack was, indeed, delivered by the French fire-ships. It was as unsuccessful as the first; but Wolfe was still utterly at a loss to devise a means of successfully closing with his enemy. Mortified at the failure of his efforts—for he was sanguine and ambitious—Wolfe now resolved to take the bull by the horns, and to deliver a frontal attack upon Montcalm's camp. After detailing an adequate force to hold Point Lévis and his entrenchments on the

[1] On the 25th a report reached the camp that the Indians in the enemy's service had burned alive three soldiers of the Royal Americans. Wolfe sent in a flag of truce with the threat of putting three hundred prisoners to death by way of reprisal. Even among the English and French the practice of scalping was prevalent. Wolfe forbade it except in the case of Indians, or Canadians dressed as Indians.

Montmorenci, he could spare little more than five thousand men to attack a force of twice that strength in a very formidable position; but nevertheless he determined to make the venture. A mile westward of the falls of the Montmorenci is a strand about two hundred yards broad at high water, and about eight hundred yards wide at low tide, lying between that river and the foot of the cliffs. Upon this strand, which was commanded (though Wolfe could not see it) by musketry from the entrenchments above, the French had built redoubts above high water mark. Wolfe hoped by the capture of one of these redoubts to bring about a general action, or at any rate to gain closer knowledge of the camp itself. Moreover, below the falls of the Montmorenci was a ford, by which some at least of the troops on the river could join in the attack, though Wolfe held the Canadian Militia in such contempt that he was not afraid to fight them with inferior numbers.

At dawn on July 31, H.M.S. *Centurion* stood in close to the Montmorenci, followed by two armed transports; and all three dropping their anchors opened fire upon the French redoubts. At the same time the batteries on the other side of the Montmorenci began to cannonade heavily the flank of the entrenchments. At eleven o'clock a number of boats filled with troops rowed across from Point Lévis and hovered about the river off Beauport, as if meditating attack upon that point. Hour after hour passed without incident; and Montcalm, with full leisure to consider the situation, saw where the danger really lay, and concentrated twelve thousand men between Beauport and the Montmorenci. At last, at half-past five, the British batteries, afloat and ashore, opened fire with redoubled fury; and the boats made a dart for the beach. Many of them grounded upon an unknown ledge some way short of the strand, which caused some confusion and delay; but ultimately all reached the appointed spot, and the men landed with little loss. Thirteen companies of grenadiers, including of course two of the Royal Americans, led the way, with 200 men of the 2nd Battalion of the Regiment close behind them; and the 15th and Fraser's Highlanders followed in support.

Our 2nd Battalion must have been short of officers, for Ochterlony, who was in command of the 200 men just mentioned, was acting as a major although his promotion to the rank of captain was so recent that he could hardly have been aware of it. He had been wounded by a German officer in a duel the day previous, but notwithstanding his friends' entreaties, was present on parade, declaring that a wound received in a private encounter should not stop him from doing his duty to his country. Whether Ochterlony felt animosity against Captain Wetterstroom as a German is not known; but he called out to that officer, who commanded the companies of Royal American grenadiers on the immediate right of his own detachment: 'Though my men are not grenadiers, you will see we shall be first at the redoubt.' Fired by the remark, the two R.A.R. grenadier companies dashed to the front, and the remainder of the grenadier battalion with them. Despite all efforts of their officers to control the men, they rushed forward in the greatest disorder and confusion upon the nearest redoubt, without formed support, and drove the French from it. Then instantly found themselves under a tremendous fire from the entrenchments above, and for a moment recoiled. Then recovering themselves, the men made a second mad rush up the steep, slippery grass of the ascent, but only to roll down in scores, killed or wounded by the French musketry. The affair was wearing an ugly aspect, when a summer's storm suddenly broke over the combatants, effectually stopping the fire by drenching the ammunition of both sides, and checking all further attempt at assault by making the grassy slopes more slippery than ever. Wolfe, seeing that everything had gone wrong, seized the moment to order a retreat; and the troops fell back and re-embarked.

Meanwhile Ochterlony, while leading his men up to the enemy's entrenchments had received a musket ball in the lungs; but though obliged to drop the fusil which his previous wound barely permitted him to carry [1] he contrived to advance until forced to stop by loss of blood. About the same time his brother officer, Ensign Peyton, was wounded by a shot which shattered the small bone of his left leg.

After the failure of the attack, a party of men retiring came up to carry the wounded officers off the field. But in spite of their earnest entreaties, Ochterlony, although unarmed—for the officers carried no swords in action—refused to be taken away, and Peyton declined to abandon his captain.

From the Indians in the pay of the enemy, the best that they could expect was swift death. Two shortly approached, but as they were accompanied by a French soldier, Ochterlony, a good French scholar, expressed to the latter his conviction that his friend and himself would receive the treatment due to officers as prisoners of war. The Frenchman's only reply was to snatch Peyton's laced hat from his head and rob the captain of his watch and money. The Indians at once accepted the outrage as a signal for murder. One of them clubbing his musket, struck at Ochterlony from behind, while

[1] The officers of the Royal Americans habitually carried fusils (or rifles) as their principal weapon.

the other shot him in the chest; and then springing upon the defenceless officer, stabbed him in the abdomen with his scalping knife; but finding him still alive, the ruffians tried to strangle him with his sash. It must have been the affair of a moment; but Peyton, having a double-barrelled gun, shot one of the Indians dead. The second, leaving Ochterlony, rushed upon his new antagonist. Each fired at the other without disabling him; but the savage continuing his rush, drove his bayonet into Peyton's body. As he repeated the blow, Peyton seized the Indian's musket with his left hand. Then, drawing a knife from the man's belt with his right, he succeeded, after a violent struggle, in despatching him. Turning over the body Peyton found that although his bullet had not stopped the savage it had pierced his chest.

He now looked round for Ochterlony and saw him sixty yards off attended by the French soldier, who appeared to have taken compassion on him. Shouting a farewell, for he saw a party of Indians approaching and thought his fate was sealed, Peyton seized his musket; and notwithstanding his broken bone, hobbled along for about forty yards till exhausted. Then he turned at bay. The two foremost Indians halted, but the French from their breastworks continued to fire at the wounded officer. Happily Peyton noticed at this moment a Highland officer with a party of men, and signalled for help. Three British soldiers came up and carried him to a place of safety. Captain Ochterlony was taken into Quebec where he died a few days afterwards. The French surgeons who attended him declared that he would probably have recovered from the two shots which he had received in the chest, and that it was the Indian scalping knife which caused the mortal wound. General Townshend afterwards made a vigorous protest against the inhuman conduct of the enemy. The French commander in reply, made the feeble excuse that the fire was kept up, not by the regular troops, but by the Canadian Militia and savages, whom they were unable to restrain.

In this action, the grenadiers and 60th had lost 500 officers and men, killed and wounded, or nearly half their number, and the casualties among the officers numbered over thirty. Many of the wounded were scalped by the Indians. Other details as to the casualties of the Regiment upon this occasion are unfortunately wanting, Wolfe having forwarded no list except a general one, which included all losses up to September 2. Since, however, in this list the casualties of the 2nd Battalion amount to 10 officers and 110 men killed and wounded, it seems probable that the detachment of the battalion which was engaged in this action suffered very severely.[1]

Had Wolfe been so unfortunate as to live in the twentieth century, the clamour of the Press and of members in both Houses of Parliament would probably have forced the Government to supersede him. As things were, he went quietly to his headquarters, and issued a stinging general order in condemnation of the grenadiers, and in commendation of the 15th and the Highlanders who had shown order and discipline. 'Such impetuous, irregular and unsoldierlike proceedings,' he said of the grenadiers, 'destroy all order, and put it out of the General's power to execute his plan. The grenadiers

[1] A casualty list supplied by Major-General Terry gives the following casualties of the 60th in this action. *Killed*: Captain Ochterlony (died of wounds); Lieutenants Kennedy and De Witt; Ensign Johnston; 1 N.C.O.; 18 privates. *Wounded*: Major Prevost; Captain-Lieutenant Brigstock; Lieutenants Ecuyier, Grandidier, Archbold, Howarth; Ensign Peyton; 7 N.C.O.s; 114 privates. This, however, does not agree with Wolfe's return; and Prevost as we have seen was wounded in a skirmish ten days before the action.

could not suppose,' he continued, 'that they alone could beat the French army.'[1] But unfortunately this was precisely what the grenadiers did suppose on this occasion, and what other corps have supposed since then on similar occasions, with the inevitable consequence of spoiling the General's combination. 'However,' concluded Wolfe, 'the loss is very inconsiderable, and may be easily repaired when a favourable opportunity offers, if the men will show a proper attention to their officers.' And, there being no special correspondents present, the whole incident was therewith very properly closed.

Much elated by their success the French now flattered themselves that the campaign was practically ended, and in their favour. Wolfe resumed his tactics of laying waste the country round Quebec, but Montcalm rightly refused to be enticed from his fortified lines, and Canada was not to be won by the burning of a few villages. Wolfe, therefore, once again took advantage of the enterprise of the fleet to resume his operations above and on the reverse side of Quebec. With every fair wind more and more ships had run the gauntlet of the batteries unharmed; and on August 5 a flotilla of gun-boats safely accomplished the same passage, while 1200 men under Brigadier Murray marched up the south bank of the St. Lawrence to do service in them. This movement obliged Montcalm to detach 1500 men from the camp at Beauport, and thus prevent any attempt at landing above the city. So vigilant however was the

[1] *Cf. Lord Wellington's G.O. of May* 1811: 'The officers of the Army may depend upon it that the enemy to whom they are opposed is not less prudent than powerful. Notwithstanding what has been printed in gazettes and newspapers, we have never seen small bodies unsupported successfully opposed to large; nor has the experience of any officer realised the stories which all have read, of whole armies being driven by a handful of light infantry and dragoons.'

small body of men, commanded by Bougainville, A.D.C. to Montcalm, in its duty of guarding fifteen or twenty miles of the river, that Murray could effect little, and only after two repulses succeeded in burning a large magazine of the enemy's stores.

In Wolfe's General Order of August 9 we read: 'The two Grenadier companies of ye Royal American battallions are to embark in four flat bottomed boats, to fall down with ye tide and escort ye General at 6 A.M. to St. Joachim.' This perhaps was the occasion mentioned by Patrick Murray (p. 292) on which the grenadiers of the two battalions were employed in clearing the country below the falls from the incursions of Canadians and Indians led on by a priest[1] who was killed in one of the skirmishes. The records of the 1st Battalion state: 'In executing this service, these companies displayed so much alertness and intrepidity on all occasions, and General Wolfe being so much pleased with their spirited conduct conferred on them the motto of "Celer et Audax."'[2]

In spite of everything the French began to grow uneasy. All the supplies of Quebec were drawn from districts farther up the St. Lawrence, and were sent down by water from want of land transport. The British fleet now effectually closed the river; there was no hostile squadron to combat it; and the only hope for the French was that winter would drive the British ships to sea before the city was starved out. Amherst also had successfully begun his advance by Lake Champlain, and had taken both Ticonderoga and Crown Point. Montcalm felt anxious, but his discouragement was as nothing to that of Wolfe whose health, always weak, gave way altogether on August 20 from anxiety and disappointment, and compelled him to delegate the conduct of operations to his

[1] M. Robineau. [2] *Vide* Appendix I.

brigadiers. Several plans were laid before them, all of which they rejected in favour of an attempt to gain a footing on the ridge above Quebec, cut off Montcalm's supplies completely, and compel him to fight or surrender. The difficulty and hazard of the operation were great; but Wolfe, on resuming command on August 31, accepted it in desperation without demur. Since the end of June he had lost over eight hundred and fifty, killed and wounded, and a still greater number of sick men; and the two battalions of the 60th alone returned 4 officers and 121 men as wounded. Supplies were scarce; Hyson tea we are told was £1 10*s.* per lb. To go on as hitherto was impossible. The co-operation of Amherst had been expected but Amherst came not, and on September 2 Wolfe wrote to Pitt the report of his proceedings in a tone of the utmost dejection.

On the following day (September 3) the troops were withdrawn skilfully and without loss from the camp of the Montmorenci; for although Montcalm threatened to attack the rear-guard he was stopped by the action on the part of General Monckton who made a feint of landing at Beauport. On the 4th a flotilla of flat boats succeeded in passing above the town with baggage and stores. On the 5th, six battalions, including the Light Infantry and the grenadiers of our 2nd Battalion, marched westward overland from Point Lévis, and embarked with Wolfe himself in Admiral Holmes' ships. Montcalm therefore reinforced Bougainville to a strength of 3000 men, and enjoined upon him the need for extreme vigilance. Bougainville's headquarters were at Cap Rouge; with fortified posts at Sillery, six miles below it, at Samos, still farther down, and at Anse du Foulon, now called Wolfe's Cove, a mile and a half above Quebec. At this last place Wolfe, while searching the heights with his glass, noticed a narrow path winding up the face of the precipice, with a cluster of only ten or twelve tents above it. He concluded therefore, that the guard at that point was small and might easily be overpowered. The weak point, such as it was, long desired, was at last found.

On the morning of September 7 Holmes weighed anchor and sailed up to Cap Rouge. The French at once manned their entrenchments, and Wolfe, ordering his men into the boats, kept them rowing up and down as if in search of a landing-place. During the four following days these feints were repeated; the ships drifting up to Cap Rouge with the flood-tide, and dropping down with the ebb, until Bougainville fairly wore his troops out with incessant marching to and fro. Finally on the 12th, two deserters came in from the camp of that officer with the news that a convoy of provisions would pass down the river to Quebec by the next ebb tide; and Wolfe saw his opportunity was come.

He sent orders to his standing camp that every man who could possibly be spared should march at nightfall up the southern bank of the river, and halt at an appointed spot for embarkation. As the dusk fell, Admiral Saunders moved out of the basin of Orleans with his ships and ranged them along the whole length of the camp of Beauport. Then, manning every boat in the fleet he opened a heavy cannonade as if to clear the beach for a disembarkation. Montcalm, completely deceived, massed the whole of his troops at Beauport, and kept them under arms. Meanwhile Holmes' squadron remained anchored off Cap Rouge, showing no sign of activity until dusk, when seventeen hundred men, the Light Infantry among them, entered the boats and drifted upwards with the flood tide, as if threatening Bougainville's headquarters.

At two o'clock in the morning the tide ebbed. Wolfe and his staff stepped into their boats, and with the 1700 men above-named, dropped silently down the river. After due interval followed a detachment of sloops and frigates, and then Holmes' ships with a second division of 1900 men, 400 of them being of the Royal Americans. Bougainville saw them go, and resolved not to weary his men with marching after them. After two hours' drifting the flotilla was challenged by a sentry; but an answer in French was returned by a Highland officer, who had been in the French service; and the boats were allowed to pass. A second soldier at Samos was similarly outwitted; and, when at last the leading division rounded the point of Anse du Foulon, not a sentry was to be seen. The troops quietly disembarked and the light was just breaking as twenty-four volunteers of the Light Infantry under Colonel Howe (afterwards the Sir William Howe who commanded in chief during the earlier phases of the American war) began to make their way through the stunted bush that covered the face of the precipitous cliff. Arrived at the top they overpowered the guard, though not without some firing; whereupon Wolfe, who had not meant to risk more than 150 or 200 men, having gained a foothold, gave the order for the remainder to advance, and the troops swarmed up as best they could. The zig-zag path which Wolfe had noticed was found to be obstructed by trenches and abattis, but these were quickly cleared away. Presently the sound of cannon was heard up the river, the batteries of Sillery and Samos having opened fire upon the second division of boats. Howe and the Light Infantry marched off at once to silence them, and effectually accomplished their purpose.

The light company of our 2nd Battalion, commanded by Lieutenant Daniel McAlpin, climbed a wooded precipice, drove out of the 'Bishop's House' a garrison, one hundred strong, and routed a detachment of equal strength sent from Sillery to retake it. Tradition says that poor McAlpin ended his days in a workhouse at New York. It is, unfortunately, not only 'to the grave' that the path of glory may lead.

Meanwhile the boats, as fast as they were emptied, rowed back to land the men of the second division from the ships. The detachment sent from the standing camp on the opposite shore was also conveyed across. The sun was still low when the whole force had ascended the cliff and was filing across the tableland which ends in the promontory of Quebec.

About a mile eastward of the landing-place was a tract of grass known as the Plain of Abraham,[1] where the plateau is narrowed to a breadth of half a mile between the cliffs of the St. Lawrence on the south and the heights overlooking the St. Charles on the north. The ground was fairly level, though broken by clumps of scrub and patches of corn; and here Wolfe decided to give battle. His entire force did not exceed, and may have been something less than, 4800 officers and men. Of these, 2850, comprising seven battalions, formed in triple rank,[2] were posted fronting due east. On the right, by the brink of the cliffs of the St. Lawrence, was a single platoon of the 28th; and then, in succession, the 35th, the grenadier companies of the 22nd, 40th, and 45th, termed the Louisbourg Grenadiers;[3] the remainder of the 28th, the 43rd, 47th, Fraser's Highlanders, and 58th. Straight through the centre of the position ran the road from

[1] Named after Abraham Martin, a pilot, familiarly known as Maître Abraham.

[2] *Vide* Townshend's despatch.

[3] Patrick Murray states that the grenadiers of our 2nd Battalion were also included in the Louisbourg Grenadiers and took the left of their line.

Sillery to Quebec, and on it was posted a single light field-gun which had been dragged up from Wolfe's Cove. On the left of the 58th, which did not reach the heights above the St. Charles, the 15th Foot was thrown back in protection of the left flank. As a reserve in rear of the fighting line were stationed the 2nd Battalion of the Royal Americans on the left (322 of all ranks [1]) and the 48th on the right, each distributed into four subdivisions with wide intervals. Two companies of the 58th guarded the landing-place ; the 3rd Battalion of the Royal Americans (540 of all ranks [1]) formed the connecting link between them and the line of battle ; and far in the rear Howe, with the Light Infantry, occupied a wood from which he could keep Bougainville in check. Monckton, of the 60th, commanded the right of the fighting line under Wolfe in person ; Murray the left ; and Townshend took charge of the 15th and 2nd Battalion of the R.A.R.

Meanwhile Montcalm, perturbed by the guns of Saunders and the threatening of Beauport by the boats of the fleet, had been awake all night. At 6 A.M., hearing the sound of cannon above the town, he mounted his horse, and to his horror perceived the scarlet uniforms on the Heights of Abraham. Realising at a glance the imminence of the peril, the General sent an aide-de-camp in hot haste to summon the troops from the right and left of Beauport camp. The men, only just dismissed from a long night's watching, streamed hurriedly through the city, partaking of the nervous uneasiness which their General had communicated to them. Then there was fresh confusion. The Governor of Quebec, M. de Vaudreuil, was not to be found ; the commandant of the city declined

[1] Another return gives the following: 2nd Battalion (probably minus the grenadier company)—2 Captains, 6 Lieutenants, 6 Ensigns, 21 Sergeants, 10 Drummers, 218 Rank and File ; 3rd Battalion—1 Lieutenant-Colonel, 4 Captains, 11 Lieutenants, 11 Ensigns, 28 Sergeants, 4 Drummers, 474 Rank and File.

to comply with Montcalm's request for twenty-five field-guns; and the division from the left of Beauport camp for some reason failed to come up. But the day was young; two or three hours would have sufficed to bring up the missing troops, to get guns enough to blow Wolfe's force off the face of the earth, and to concert measures with Bougainville. It was impossible that the British could have brought any quantity of supplies with them—as a matter of fact they had provisions for only two days—and with his superior numbers and greater wealth of sharpshooters Montcalm could have worried them all day and all night, and might not impossibly have hindered the bringing up of fresh supplies, in which case Wolfe's position would have been critical. As it was the Marquis held a hurried council of war, and decided to accept battle at once with the five thousand men that were with him.

By nine o'clock his line of battle was formed about six hundred yards from the British position. On his right and left were two battalions of Canadian Militia; and between them, formed in columns, five regular battalions drawn from the grand regiments of La Sarre, Languedoc, Royal Roussillon, and Béarn. Wide upon both flanks a swarm of two thousand Indians and sharpshooters advanced in skirmishing order, finding plenty of cover not only on the flanks but also in the patches of scrub in front. Wolfe threw out skirmishers to meet them, and the fire became lively, especially on the British left, where the 15th began to fall fast. Alarmed for this flank, Townshend called up the 2nd Battalion of the 60th to the left of the 15th, using part of it to drive the Canadians from some houses by the Sillery road and doubling back the remainder *en potence* in line with the 15th. Hard fighting took place. At the same time the

Light Infantry was summoned also to support the same point, and thus the rear-guard and half of the reserve were absorbed into the fighting line before the action was well begun. Happily the fire of the enemy's skirmishers on the British right, where they could not get round the flank, was by no means so deadly as on the left; and along the greater part of the front at any rate Wolfe made his men lie down.

Montcalm meanwhile had brought up three field-guns on the Sillery road, which opened fire and were answered by Wolfe's solitary gun with great effect. At last about ten o'clock the whole French line advanced with loud shouts to the main attack, and the British sprang to their feet and stood steady with recovered arms. Arrived within two hundred yards, the French immediately opened fire, with little effect though Wolfe was shot through the wrist; and the Canadian Militia caused much confusion and delay by throwing themselves flat on the ground (like good skirmishers) to reload. Having recovered their order the French again came on, but the British stood motionless to let them come within close range. In those days great value was attached to the first volley in an action, since it was the only one for which the officers could be certain that the muskets were properly loaded. On this occasion Wolfe had given orders to load with two bullets; and, like Wellington half a century later, allowed the French to approach within thirty-five yards before he gave the word to fire. Then the British, as if on parade, brought up their muskets, and with one crash poured in the most perfect volley ever fired on a battle-field. Instantly the French were veiled from their sight by the dense cloud of smoke. Behind the cloud was the sound of a confused hubbub, while the British reloaded and stood fast until the smoke rolled away. Then the French line was seen to be shivered to pieces, and the ground covered with the fallen; while an officer on a black charger, Montcalm himself, was galloping frantically up and down the shattered ranks in a vain attempt to restore order.

Again Wolfe gave the order to fire, and then the whole line advanced with bayonet and claymore. There was nothing to hinder them except a few sharpshooters who kept up a galling fire from the scrub and caused heavy loss before they were dislodged; but when these were dispersed the whole French force broke up and fled in wild confusion. Wolfe, standing near the Louisbourg Grenadiers, had been struck in the groin and breast, and carried to the rear to die;[1] Monckton also had been shot through the body; and it fell to Townshend to rally the pursuing troops and meet a new enemy, namely Bougainville,[2] who now appeared on the scene. Approaching the left rear of the British, Bougainville's position was precisely that of Blucher at Waterloo. But confronted by our 2nd Battalion, Howe's Light Infantry, the 15th, and two field-guns (the second of them apparently captured from the French) Bougainville declined to drive home the attack. He retired; yet the Canadian Militia under cover of the brushwood opened a galling fire upon our left and checked the pursuit until regularly attacked and driven away. Then Townshend, having improved the road up the face of the cliff for the transport of cannon

[1] Some writers say that Wolfe fell into the arms of Lieutenant Brown of the grenadiers. This may have been Lieutenant John Brown of the Royal Americans. On the other hand, Morgan, in his work *Biographies of Celebrated Canadians*, states definitely that he fell into the arms of Lieutenant Des Barres of our Regiment, who was in the act of making a report.

[2] Bougainville, distinguished as a soldier, sailor, botanist, and explorer, after a life of adventure, became a senator during the Consulate, took part in Bonaparte's great work of constructive statesmanship, and lived to see the Empire in the height of its glory. He died in August 1811. The Bougainvillæa plant commemorates his name.

and supplies, entrenched himself on the battle-field. No rest could he give his troops until after midnight, by which time they had been under arms for over thirty hours.

In this action the losses of the British were 9 officers and 48 men killed; 36 officers and 535 men wounded. Of these, 6 men killed, and 5 officers and 83 men wounded—over thirty per cent. of its strength—belonged to the 2nd Battalion of the 60th; the 3rd Battalion, which had hardly been engaged, having two men wounded only. The loss of the French was at least 1500 in killed, wounded, and prisoners; and one of their three guns was captured.

After the action, confusion and strife reigned within the French camp. Montcalm—than whom our Regiment has never encountered a more gallant antagonist—wounded during the victorious advance of the British, was dying; the civil governor—the boastful Vaudreuil—was timid and irresolute. The French were still far superior in number to the British, but their leaders were distracted; and when Vaudreuil at last gave the order to retreat, the entire French force, excepting a small garrison which was retained for Quebec, streamed away during the night in disorderly flight up the St. Charles. Townshend on the days following the action pushed his trenches forward against Quebec with the greatest energy, and on the 17th, at the sight of a mere advance of red coats, the city surrendered. The articles of capitulation had hardly been signed when intelligence came that a French force, under M. de Lévis, second-in-command to Montcalm, was about to march to raise the siege; but before de Lévis was well on his way he learned that he was too late. In the afternoon the British occupied the city, and spent the following weeks in improving the defences and making provision against the winter.

Thus ended a series of operations memorable in many ways, and to us especially memorable in that it gave to our Regiment—from Wolfe himself—the motto 'Celer et Audax.'

Brigadier-General Murray was left in command with a garrison of 7500 men, including the two battalions of the Royal Americans. At the end of October Saunders fired his farewell salute, and sailed with his fleet to England, carrying with him the body of Wolfe to its final resting-place in the parish church of Greenwich.

The connection of the 60th with Quebec did not entirely terminate until November 1871, when a detachment of the same battalion which had fought at the Plain of Abraham hauled down, on quitting the city, the Imperial flag which was then replaced by that of the Dominion of Canada.

We must now turn to the operations of Amherst, and of the subsidiary columns which were acting under his direction. The General's plans were directed upon three different points. One column was to advance by way of the Little Lakes upon Montreal; another was to seize Niagara; a third to relieve Pittsburg which was threatened by the enemy, to capture the forts which barred the road to Lake Erie, and then push on to Niagara and release Prideaux for operations on the St. Lawrence.

Amherst in person took command of the first, about 11,000 strong, half regular troops, half provincials. Before the end of June he had concentrated his force at the head of Lake George, and was busy training it in drill, musketry, wood-fighting, and the work of scouts.

A The Fort.
B The Redoubt.
b Lower Battery of Two Guns.
C Wharf and Harbour for the Vessels of War.
c Store House.
D Earth Works.
E Stone Work.
F Well House.
G Nine Ovens.
H Provision Store House.
I Old French Battery.
K The French Lines.
L Abattis.
M Batteries.

The line of communication was protected by the erection of forts at intervals of three or four miles as far south as Fort Edward. On July 21 the army embarked on the waters of Lake George and landed next day without much opposition near the saw-mill at Ticonderoga. Here Bourlamaque was in command of 3000 men, but making no attempt to defend the position the scene of last year's victory, wisely retired under cover of the guns of the fort down Lake Champlain to Isle aux Noix. The fort was evacuated and blown up on the 26th. Crown Point was abandoned on August 1, and the way to help Wolfe seemed clear. But the French had four small armed vessels on Lake Champlain, and it became necessary to construct a naval force to defeat them. The contingency might have been foreseen and the vessels built during the previous winter. But the duty had been neglected, and by the time that the vessels had been built the French squadron had been defeated, Quebec had long been taken, and October gales stopped a further advance.

Amherst did a good deal of useful work in fortifying posts and improving communications, but he neglected the essential duty of pushing forward to co-operate with Wolfe. One must feel that a more enterprising and far-seeing commander might have established himself on the St. Lawrence before the end of August, and have effected a diversion in favour of Wolfe at a time when that General was driven to the verge of despair. 'How General Amherst will excuse himself to his Court, I don't know,' wrote one of the French generals.

Concurrently with his own advance General Amherst assigned to Brigadier Prideaux the duty of reducing Niagara, and to Brigadier Stanwix of the 60th that of relieving Fort Pitt. Prideaux assembled his force at

Shenectady on the Mohawk River, its numbers being in all 5000, among which were included the 44th, 46th, the 4th Battalion of the 60th, and 2500 men of provincial troops from New York. On June 15 he began his march up stream by Lake Oneida and the river Onandaga to Oswego, where, leaving nearly half his force under Colonel Haldimand to secure his communications, he embarked with the rest upon Lake Ontario for Niagara.

Haldimand at once set to work to erect a fort; but while his men were dispersed to fell timber the French officer La Corne at the head of a thousand men made a sudden appearance. But for the fact that Father Piquet, after the manner of his cloth, thought the occasion favourable for a lengthy discourse, and a panic among the Canadians who ran away at the critical moment, upsetting the worthy priest in their flight, the British might have been taken by surprise. As it was the Royal Americans hurriedly got under arms and the enemy was eventually driven back to his boats with the loss of thirty killed and wounded, among the latter being La Corne himself. Haldimand had but two men killed and eleven wounded.

At Niagara the fort was situated in the angle formed by the Niagara River and the lake, and garrisoned by 600 men. Prideaux at once laid siege to it in form, though his engineers were so unskillful as considerably to delay the operations. When at length the battalion did open fire Prideaux was killed by the premature explosion of a shell from one of his own guns. The command should have devolved upon Colonel Haldimand, but was claimed by Sir William Johnson, an Irish emigrant, who from his extraordinary influence over the Indians, had long been the agent for New York in all dealings with them, and had gained the honour of knighthood for his services. Sir William had seen service but did not belong to the

British army. Yet, fearing a quarrel which might spoil the whole scheme, Haldimand gave way. Johnson pressed the siege with vigour, and the fort, after two or three weeks' cannonade, was already in extremity when a party of 1300 French and Indians came suddenly on the scene, intent upon relieving the place. Johnson therefore left one-third of his men to occupy the trenches, and another third to guard the boats, while sallying forth on July 24 with the rest—fewer than a thousand men—he routed the French completely after a brisk engagement. The enemy fled in haste, burned Venango and the posts on the lake, and retired to Detroit. Niagara surrendered on the same evening, and the great object of the expeditions of Prideaux and Stanwix was thus gained. The whole region of the Upper Ohio was cleared of the French, and their posts in the west were hopelessly cut off from Canada. Amherst at once sent Brigadier-General Gage to replace Johnson, with orders to attack the French post at La Galette, by the head of the rapids of St. Lawrence, and thence to push on as close as possible to Montreal. But Colonel Haldimand felt the project was impracticable. Gage concurred, and the campaigning season came to an end without any important movement on his part.

Meanwhile General Stanwix had, with enormous difficulty, collected six weeks' provisions for 5000 men at Fort Bedford, his force being made up of the 1st Battalion[1] of the 60th, under Major Tullikens, with three battalions of Provincials. This battalion had been properly equipped for forest fighting, somewhat after the Indian fashion. 'Our people have leggins,' wrote Major Tullikens to Bouquet in March 1759. 'I intended always that they

[1] It seems that six companies only were present, and that the remaining four —weak in numbers—were at Lancaster.

should have blue, but we could not have blue at Albany, so that we have green tied with a red garter.' 'Leggins'—the usual spelling at the time—were really 'putties.'[1]

Stanwix set out on his march at the end of May or beginning of June; but with all preparations, the advance upon Fort Pitt was always difficult and arduous. On July 30 the General detached Major Tullikens with his battalion to push on through Fort Pitt and Venango to Niagara, which he was directed to garrison, and thus

[1] The following is a state of the battalion: that of the first six companies at Albany being dated February 24, and that of the remainder at Yorktown, March 24, 1759.

Companies	Capts.	Lieuts.	Ensigns	Adj.	Sergeants	Drummers	Privates	Wanting to complete
Brig.-Gen. Stanwix'	–	1	–	1	3	2	84	16
Major Tullikens'	–	1	1	–	2	1	72	29
Capt. R. Mather's	–	1	1	–	3	1	86	14
,, Gavin Cochrane's	1	1	–	–	3	1	79	21
,, W. Stewart's	1	1	–	–	2	1	84	16
,, J. Schlosser's	?	?	?	–	?	?	86	14
	2	5	2	1	13	6	491	110

Absent.

Brig.-Gen. Stanwix, at New York
Major Tullikens, at New York, health
Capt. Mather, Recruiting
Lieutenant Mayer, Recruiting
,, Barrett, Engineer with Gen. Forbes
,, Brehm, Scouting party
Lieutenant Ralfe, Recruiting
,, Wilson, in England
Ensign McIntosh, New York
J. Dow, on command, Flat Bush
Chaplain Jackson, Major-Gen. Abercromby's Staff
Surgeon Stevenson, Brig.-Gen. Stanwix' Staff.

	Capt.	Lieut.	Ensigns	Surgeons	Sergts.	Drummers	Privates	Wanting
Col. H. Bouquet	1	1	–	1	3	2	86	14
Capt. R. Harding	–	1	–	–	3	1	84	16
,, F. Lander	–	1	1	–	2	1	87	13
,, T. Jocelyn	–	–	1	–	4	2	85	15
	1	3	2	1	12	6	342	58

In March, 1 Sergeant, 1 Corporal and 12 Privates from each company were formed into a Light Company.

release a part of Prideaux' force for operations on the St. Lawrence. On reaching Fort Pitt in person with his main body, Stanwix could only report that the Ohio was so low that it was impossible for him to make his way to Niagara. 'The 1st Battalion of the 60th,' he added, 'are shot at every day'—probably by skulking Indians—'but behave well.' Fortunately Prideaux and Johnson, as already mentioned, had effectually cleared the French from every post between Pittsburg and Fort Erie, so that it was sufficient to send forward the Royal Americans to occupy Venango, Fort Le Boeuf and Presqu'île.

Colonel Bouquet had the appointment of D.A.G. to this column; but his duties, as shown by his correspondence, were those of officer in charge of the line of communications. The work of procuring transport and forwarding supplies was extremely difficult, and the convoys were constantly attacked on the march. In a letter dated September 13 he writes:

> 'We have had an account from St. Lawrence that General Wolfe attacked the lines on July 31 with all the Grenadiers and 200 R.A. [Royal Americans—this was the premature attack at the Montmorenci falls], but was repulsed with the loss of 400 men partly wounded. People begin to think he will not succeed but will ruin the country in his Retreat. I hope better; he has beat the Canadians and Indians everywhere, killed great numbers and got 500 prisoners.'

Bouquet little knew that Wolfe had fallen in the moment of victory on the very date he was writing. Rumours of that victory did not reach him until October 24.

It then remained to fix the winter quarters of the troop. Of the 1st Battalion of the 60th, five (or six) companies were left at Pittsburg, and four (or five) at Lancaster. Of the 4th Battalion, nine companies were at Fort Ontario or Oswego, and one company at the falls of the Onandaga; while the 2nd and 3rd Battalions had each three companies in Nova Scotia, and seven in Quebec.

Canada still remained in the hands of France; but in letters of fire the handwriting was on the wall, and to any person of reflection the end was evidently but a question of months.

CHAPTER IV

OCCUPATION OF QUEBEC—ADVANCE OF LEVIS—BATTLE OF SAINTE FOY—RELIEF OF THE CITY—AMHERST ADVANCES AGAINST AND TAKES MONTREAL

WOLFE was dead: Monckton badly wounded. Townshend had sailed home with the fleet to blow his own trumpet in England.[1] Murray remained to command the garrison of Quebec. Officers of our Regiment who have served at that city have described the intense cold of the place; and it was with gloomy feelings that the British troops looked forward to passing the winter there. It is almost needless to state that Lord Barrington, the British Secretary of State, had neglected to forward the warm clothing essential for protection against the rigour of snow and frost. Working parties laboured incessantly for weeks to cut down trees even at a distance of four or five miles for the provision of fuel which even so might prove inadequate. The town was in ruins, and the quarters were necessarily rough. But matters might have been worse. Murray kept good discipline although, being young and inexperienced, his method was Draconian. Men were awarded such punishments as a thousand

[1] In later years Townshend became Colonel of the 2nd Dragoon Guards, and wrote a book called *An Academy for Young Horsemen.* The celebrated work of 'Geoffrey Gambado,' so often alluded to by Mr. Jorrocks in *Handley Cross*—a work which every Rifle officer should know by heart—was written as a skit on Townshend's book.

lashes; mitigated to three hundred in consequence of the severity of the weather.[1] Horrible to relate, even two women were flogged for selling liquor. As regards the relations between French and English, conquerors and conquered got on well together. The General Hospital, with French nuns as nurses, was admirably conducted.

Towards the end of November a French squadron laden with supplies attempted to sail past the batteries, and the majority of the vessels seem to have succeeded. The old year passed away, and the new year—1760—was ushered in with terrible suffering. In February parties of the enemy appeared and skirmished unsuccessfully with the outposts. A party under Major Dalling, equipped with snow-shoes, crossed the St. Lawrence on the ice, drove off the enemy in rout, and a few days later repulsed a somewhat formidable counter-attack. Other episodes of petty warfare took place, ending for the most part in our favour.

During all this time rumours of an attack in force were rife. But April came; the ice began to melt and the alarms were not verified. On the 27th however, long before daybreak, the watch on board H.M.S. *Racehorse* (Captain Macartney) reported the cry of a man apparently in distress. In spite of the darkness and the danger from the masses of ice floating down the stream, a boat was launched; and, wonderful to relate, came upon a large floe on which was lying a man all but dead with cold. When revived he stated that he was an artillery sergeant of General Lévis's army, 12,000 strong, which was approaching with the intention of recapturing Quebec. General Murray was instantly informed and got the garrison under

[1] In Cromwell's army, the maximum number of lashes inflicted was sixty. It is difficult to understand how a century later, with presumably an increase of civilisation, such fearful barbarities as here described could take place. Yet the code was still in force in the time of Wellington.

arms. Yet had it not been for the threefold providence which saved this French sergeant alone of his boat-load; which, having carried him on his block of ice far below Quebec upon the ebb tide, brought him back on the flood to the precise spot where the *Racehorse* lay at anchor; and which enabled the boat's crew to find him in the darkness, Murray would have had no warning-note of his imminent peril.

For lack of transport and other reasons the long-expected attack had been deferred. It had been quietly prepared with all the foresight of a resolute and experienced commander. Naval co-operation was essential; and under command of a fine sailor, Captain Vauquelin, a squadron of two frigates, two sloops of war, and two schooners were still available. The field-guns having been hoisted on board these vessels, and the infantry embarked in large boats, Lévis set out from Montreal on April 20; his military force including eight battalions of the line and two colonial, 3000 Canadians and 400 Indians. The distance by river from Montreal to Quebec is about 170 miles. Descending the river, he picked up *en route* the garrisons of Jacques Cartier, Deschambault, and Pointe aux Trembles, with the Canadian Militia on each bank. The French sergeant's report was an exaggeration; but with these additions, the French army was little, if at all, less than 9000 strong. On the 26th it landed at St. Augustin, crossed the Cap Rouge River, and drove in the British outpost at Old Lorette. It was now late in the day, but despite a terrible storm of wind, rain and thunder, Lévis continued to advance during the night. It seems incredible that the force was able to make any progress at all. The route ran through a swamp, the roads and causeways over which broke under the tramp of marching feet. Amid pelting rain and furious gusts of wind, nothing but the vivid flashes of lightning enabled the troops to know that they were on the right track. The way was, no doubt, familiar: yet the feat was a marvellous one. On they marched despite the difficulties of ice and bitter wind; and at daybreak, not two hours after Murray's fortunate warning, found themselves close to Sainte Foy, where the army was brought to a halt by the fire of a British post occupying the village.

About the same time General Murray was starting from Quebec with a view to withdrawing his outposts at Sainte Foy, Sillery, &c. He effected his object, and returned to the city undecided whether to fight in the open or behind its ramparts. Prudence seemed to dictate the latter course; for, in default of fresh meat, such ravages had scurvy made in his garrison that nearly 700 men had died, and, of the remainder, only 3000 were fit for duty, if indeed 'scorbutic skeletons,'—to use the descriptive term of an eye-witness, can be considered to come under that heading.[1] On the other hand the walls of Quebec had not been repaired after the battering they had received from Wolfe: a large force of field artillery was available; and from the result of the battle of the previous September and of many more recent combats his men, despite their physical weakness, were full of confidence.

To those of us who know Murray from his portrait when a wizened, toothless old man, it is perhaps difficult to imagine him at this period, in the flower of youth, full of dash and ardour. But such was the case, and he decided to give battle outside the walls. At 6.30 A.M. he accordingly marched from the city with his 3000 men,

[1] The bodies of these 700 men were awaiting the thaw for interment! It is curious that no single woman died, and that hardly any contracted illness.

two or three howitzers and twenty field-guns. The ground was still frozen; but the snow and ice had partially thawed.

In his subsequent despatch to Pitt the General says that he was resolved to give battle; but the number of picks and shovels which he took gives the impression that it would have been more correct to say that he meant to receive battle in an entrenched position. If so, circumstances changed his intentions. On reaching the position occupied by Montcalm the year before, Murray halted to reconnoitre, and was at once able to observe the enemy advancing rapidly from Sainte Foy. Two French brigades were in fact already deployed on the plateau near the Anse de Foulon, and formed the right of Lévis' line. The remainder of his troops were following, but were not as yet in position, and his left was *en l'air*. Murray thought he saw his chance of defeating the enemy in detail and at once advanced to the attack. His force was formed in three brigades. That on the right, commanded by Colonel Burton, comprised the 15th, 2nd Battalion of the Royal Americans, and the 48th. That on the left, the 43rd, 47th, Fraser's Highlanders, and 28th, under Colonel Fraser. The third brigade, consisting of the 35th and 3rd Battalion Royal Americans, under Colonel Young was held in reserve. The flanks of the whole were covered by light infantry. The two battalions of the 60th were exceptionally weak. Their state on April 24 had shown 237 men fit, and 163 unfit for duty in the 2nd Battalion; 253 men fit, and 215 unfit in the 3rd. The 2nd Battalion had evidently not recovered its losses at the Montmorenci Falls and the Plain of Abraham. The reason of the weakness of the 3rd Battalion is less evident. Each battalion had still three companies—250 men—in Nova Scotia.

The guns were dragged by hand. The practice of massing artillery was as yet unknown, and the field-pieces were distributed in pairs between the battalions of the first line. On reaching the ground occupied last year by Wolfe's army, the guns unlimbered and opened so effective a fire upon the enemy's left wing, which was still on the march, that Lévis ordered his whole force to retire to the cover of the woods. The movement was executed in some disorder, and Murray advanced still farther. Then Burton's Brigade, with its cover of light infantry, got into low and marshy ground. The light troops captured the house and windmill of Dupont, and dashed forward in pursuit of its occupants. They soon fell into disorder. The guns in their present position were silenced; and a strong counter-attack on the part of the French left, which had rallied in the woods, drove back in confusion the pursuers, who in their retreat masked the fire of Burton's Brigade. The house and mill were retaken; but the French were again driven out and the buildings were left unoccupied by either side. Despite the odds a desperate struggle was maintained: but after an hour the ammunition for the guns was exhausted; the wagons had stuck fast in the snow-drifts.

On the British left, which had also met with some initial success, the form of the woods enabled the Canadians, covered by the trees, to take the left of Fraser's Brigade in flank. By degrees the right became also outflanked. The guns perforce ceased firing, and the infantry was exposed to the well-directed cannonade of Lévis' small battery of three pieces. Its losses were terrible. Young's reserve was called up. But in vain the 3rd Battalion with the 43rd charged and re-took two redoubts. In vain did the men show the steadfastness of veteran soldiers. Outnumbered and outflanked on both sides, the order was given to fall back. 'D——n it,' cried the enraged men, 'what is falling back but retreating?' But if Quebec was to be saved, retreat was the sole resource.

The retirement was conducted by alternate brigades, beginning with the left. It was found impossible to carry off the guns; they were spiked and abandoned. The retreat was effected in fairly good order and the gates of Quebec were regained in safety. The fight, although lasting but two hours, had been sanguinary. Murray's loss was close on one thousand—about thirty-three per cent. of his force. Part of our wounded were left behind, and many of them were scalped and murdered by the Indians, all of whom are said to have been Christians! The casualties in the 2nd Battalion of the Royal Americans amounted to 2 men killed, 2 officers and 9 men wounded; in the 3rd Battalion 10 men were killed, 11 officers and 32 men wounded. The French loss was estimated by the British at nearly two thousand; but Lévis' return shows only 833 casualties.

On reaching the city General Murray maintained discipline by vigorous measures. Captain S. J. Hollandt of the 60th was appointed Chief Engineer in place of McKeller who had been wounded. The gaps in the walls were filled up with fascines and sandbags; embrasures were cut and guns mounted; officers sharing the manual labour with the men. Lévis, on his side, worked with equal energy at the trenches; but Murray was beforehand with him, and overwhelmed the French with artillery fire before the latter could lay a single gun in battery.

To both commanders the best hope lay in the arrival of a friendly squadron in the St. Lawrence; the worst fear, that the squadron when it appeared might be that of the enemy. Friendly badinage was exchanged. De

Lévis offered to bet Murray £500 that he would capture Quebec. Murray declined to bet on a certainty, for he was convinced that before the year was out the French army would be sailing home in British ships! What Murray said in jest was to be shortly borne out in deed, but at the moment the prospect was as black as possible. He had now only 2400 men under arms, and they ill-fed, unhealthy, and ready to drop from fatigue. It was evident that within a few days their fate would be sealed.

On May 9 a ship of war was reported in the river. Was she friend or foe? Excitement ran high as she began to hoist her colours. But a deafening cheer from the ramparts greeted the cross of St. George, as it slowly reached the masthead and floated on the breeze. The frigate was H.M.S. *Lowestoffe*, bringing the welcome news that she was one of a squadron, the main body of which had already reached the mouth of the St. Lawrence.

Still Lévis fought on in the despairing hope that a stronger French fleet might at the last moment appear. His hopes were vain. On the 15th a ship of the line and another frigate sailed into the harbour. Next day the two frigates passed above the city to attack Vauquelin's squadron. Vauquelin fought his own ship to the last. The other Frenchmen made little resistance. Lévis saw that his last chance had gone; he raised the siege and retreated in hot haste. At daybreak on the 17th Lieutenant M'Alpin of the 60th—'a brisk, active officer,' says Captain Knox in his Journal—reconnoitring, found the trenches abandoned, and sent word to the Governor. Murray marched in pursuit. He failed to overtake the enemy, but found forty guns and all the French sick and wounded. De Lévis made good his retreat to Montreal. Vauquelin, who had been covered with wounds and captured, was sent to France, where his gallantry was rewarded by a dungeon in which he

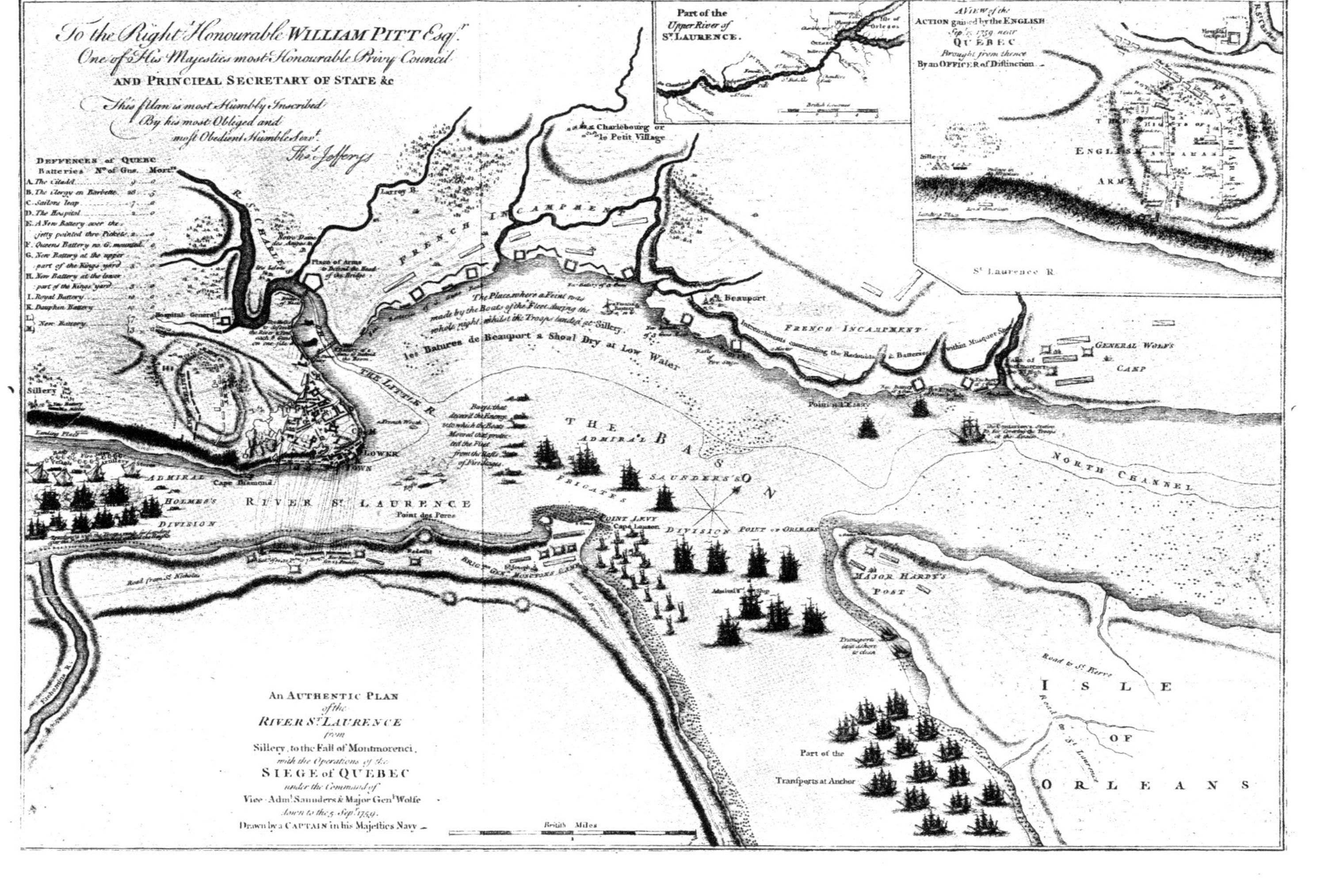

To the Right Honourable WILLIAM PITT Esq.r
One of His Majesties most Honourable Privy Council
AND PRINCIPAL SECRETARY OF STATE &c
This plan is most Humbly Inscribed
By his most Obliged and
most Obedient Humble Serv.t
Thos. Jefferys
DEFFENCES of QUEBC
Batteries N.o of Gus. Mort.rs
A. The Citadel 9 0
B. The Clergy en Barbette 28 5
C. Sailors leap 7 0
D. The Hospital 2 0
E. A New Battery over the jetty pointed thro Pickets 2 0
F. Queens Battery no G. mounted 0
G. New Battery at the upper part of the Kings yard 3 0
H. New Battery at the lower part of the Kings yard 3 0
I. Royal Battery 10 0
K. Dauphen Battery 10 0
L. M. New Battery 7 3 0
Part of the Upper River of St. LAURENCE
A VIEW of the ACTION gained by the ENGLISH Sep.r 13. 1759 near QUEBEC Brought from thence By an OFFICER of Distinction
ENGLISH ARMY
FRENCH ARMY
Sillery
St. Laurence R.
R. St. Charles
Larrey R.
Charlebourg or le Petit Village
FRENCH INCAMPMENT
Place of Arms to Defend the Head of the Bridge
Hospital General
Beauport
The Place where a Feint was made by the Boats of the Fleet during the whole night whilst the Troops landed at Sillery
les Batures de Beauport a Shoal Dry at Low Water
FRENCH INCAMPMENT
Intrenchments commanding the Redoubts & Batteries within Musquet Shot
GENERAL WOLFS CAMP
THE LITTLE R.
THE BASON
ADMIRAL SAUNDERS'S DIVISION
FRIGATES
NORTH CHANNEL
Point de Lessy
LOWER TOWN
Cape Diamond
ADMIRAL HOLMES'S DIVISION
RIVER St. LAURENCE
Point des Peres
POINT LEVY
POINT OF ORLEANS
MAJOR HARDY'S POST
Road from St. Nicholas
Road to St. Pierre
Road to St. Laurence
ISLE OF ORLEANS
Part of the Transports at Anchor
British Miles
An AUTHENTIC PLAN of the RIVER St. LAURENCE from Sillery, to the Fall of Montmorenci, with the Operations of the SIEGE of QUEBEC under the Command of Vice-Adm.l Saunders & Major Gen.l Wolfe down to the 5 Sep.r 1759.
Drawn by a CAPTAIN in his Majesties Navy

died shortly afterwards. His good name was, however, rehabilitated after his death by Louis XVI.

The honours of this little campaign undoubtedly lie with the French commander, who had conceived and carried out his enterprise with masterly skill. His energy, power of command, and ability in handling men under fire rank him high among French generals, and it was through no fault of his own that success did not in the end attend his efforts.

It is commonly believed that Wolfe's victory decided the fate of Canada. Sainte Foy shows how little that idea is justified and how precarious in reality was our tenure of Quebec. That city was saved purely by the sea-power of Britain; and so long as Montreal remained in the hands of the French, the conquest of Canada was incomplete.

Of the 1st Battalion six companies seem to have passed the early months of this year at Pittsburg and the remaining four companies—under strength—at Lancaster. It appears that a provincial force under Colonel Shippen which garrisoned Fort Bedford, showed signs of insubordination, and in an order of January 21, 1760, that officer warns his troops that the Royal Americans are marching up from Carlisle.

In the meanwhile General Amherst had been maturing his plans for the final and decisive campaign. Quebec had been the key to the casket; but Montreal was the jewel contained therein. Meanwhile the British Government had resolved to destroy and abandon the useless fortress of Louisbourg. Its garrison was consequently released for active service when wanted. The General's design was simultaneous advance upon Montreal from east, west and south. Murray was to ascend the St. Lawrence from Quebec; Brigadier Haviland was to break in by Lake Champlain. Amherst himself was to lead the main army from Lake Ontario down the St. Lawrence; his object was to cut off the retreat of the French westward, while Haviland and Murray closed in upon Montreal.

The scheme was finely conceived and justified by the circumstances of the case as well as by the result. But to the enemy holding interior lines it left the possibility of attacking in detail the assailant bodies, each of which had to traverse from a hundred to two hundred miles.

Murray was the first to move. Leaving behind 1700 men as a garrison in Quebec, and Colonel Young, termed by Knox an incomparable officer, universally esteemed, as Judge of Police, he picked out 2500 for the active work of the campaign, and embarked with them on July 14. For this occasion the 2nd Battalion of the Royal Americans and the 43rd formed a composite battalion commanded by Major Oswald; and the 3rd Battalion with the 35th Regiment formed another; the detachments from each regiment of the garrison consisting of 9 officers and 161 non-commissioned officers and men. The grenadier companies of the various regiments were formed into two battalions. The grenadiers of our 3rd Battalion were attached to the first of these battalions, those of the 2nd Battalion to the other. Captain Wetterstroom commanded the floating batteries.

Murray sailed slowly up the river, skirmishing with small parties of the enemy, and disarming the inhabitants as he passed. On August 4 he reached Les Trois Rivières; and, neglecting the French detachment which lay there, passed on to Sorel where 4000 French under MM. Bourlamaque and Dumas were entrenched upon both sides of the river. Pursuant to their orders, these

WILLIAM HAVILAND

officers followed Murray in his advance up the river; though the latter, by a mixture of gentleness and severity, persuaded half of Bourlamaque's Militia to give up its arms and remain neutral. On August 27 he reached Île Ste. Therèse, just below Montreal, and encamped to await the arrival of Amherst and Haviland. He did not wait long, for on September 3 he received, by an officer of the 60th in disguise, a message that on the 29th Haviland and his 3500 men had passed all the posts between Lake Champlain and Montreal, and were within easy march of the capital.

Vaudreuil had entertained hopes of stopping Haviland, and, to that end, had posted Bougainville, with 1700 men, three armed vessels and several gun-boats, at Isle-aux-Noix, whence he communicated with another force of 1200 men at St. John, twelve miles lower down. Haviland, who had embarked four days before at Crown Point, cleverly turned Bougainville's position, captured his flotilla and forced him to retire on St. John in order to regain his communications. The Brigadier-General following, drove Bougainville from St. John and Chambly back to the banks of the St. Lawrence, where he formed a junction with Bourlamaque.

The main army under Amherst assembled at Oswego in July, numbering about eleven thousand men, of whom 4000 were British regular troops, including the 4th Battalion of the 60th. Colonel Bouquet was left with the 1st Battalion to guard the western frontier from the Indians. On August 10, preceded by an advance-guard under Colonel Haldimand, Amherst's force embarked on a flotilla of nearly eight hundred small craft, and by the 15th had reached La Galette—the site of the present Ogdensburg. Thence, after the capture of a French brig, the flotilla pursued its way among the Thousand

Islands. At the head of the rapids stood a French post named Fort Lévis, held by Captain Pouchot, the defender of Niagara, who on being recognised, was greeted by the British soldiers with a cheer. The fort was invested and after three days' cannonade forced, on the 26th, to surrender. Four days were then spent in repairing the fort and the boats, and on the 31st the army entered upon the critical task of passing the rapids. All went well until September 4:

'Now,' says Parkman, 'began the critical part of the expedition, the descent of the rapids. The Gallops, the Rapide Plat, the Long Saut, the Coteau du Lac, were passed in succession with little loss, till they reached the Cedars, the Buisson, and the Cascades, where the reckless surges dashed and bounded in the sun, beautiful and terrible as young tigers at play. Boat after boat, borne on their foaming crests, rushed madly down the torrent. Forty-six were totally wrecked, eighteen damaged, and eighty-four men were drowned. La Corne was watching the rapids with a considerable body of Canadians; and it is difficult to see why this bold and enterprising chief allowed the army to descend undisturbed through passes so dangerous.'

On the 6th the boats glided into La Chine, about nine miles above Montreal; and Amherst disembarking his army, marched at once up to the city and invested it.

In this land defended by mountain and marsh, bristling with savage forests, divided by rivers broken with cataract and torrent, yet unspanned by bridge or ferry, the success of Amherst's combination, splendidly conceived and splendidly executed, had been remarkable. The decisive battle alone remained to be fought, but by this time the whole of the Canadian Militia had deserted the French colours, and the garrison amounted hardly to 2500 dispirited, demoralised men. Resistance was hopeless; negotiations for a capitulation were at once opened.

Vaudreuil, the Governor, drew up fifty-five articles and sent them by Bougainville to the British Commander.

General Amherst granted those which were reasonable, but flatly refused the others, including that which the French thought the most important of all, viz. the proviso that they should be allowed to march out with arms, guns, and the honours of war. In vain did, first, Vaudreuil, and then Lévis, entreat for the concession. Amherst sternly replied that by this capitulation he was determined to publish to the whole world his detestation of 'the infamous part taken by the troops of France in exciting savages to perpetrate the most horrid and unheard of barbarities in the whole course of the war, and for other open treacheries and flagrant breaches of faith.' On the 8th the French laid down their arms.

A compliment was paid to the Royal Americans in the selection of Colonel Haldimand to take possession of Montreal and act as Governor.

Under the terms of the capitulation the French troops were sent back to France after engaging to serve no more during the war. The Canadians were allowed free exercise of the rites of their (Roman) Church and full enjoyment of their property. The breadth of view which led Britain to give the utmost latitude to foreign customs, Church and legislative code in Canada, is one of the most remarkable and satisfactory features of her history. The toleration displayed seems to have been largely due to the statesmanship of our Colonel. Haldimand proved an excellent administrator, and, although the French navy continued to threaten the St. Lawrence, and five companies of the 4th Battalion Royal American Regiment were held in readiness to repel raids, it soon became evident that the mass of Canadians were happier under British rule than they had been under that of their own sovereign.

Amherst has been sometimes criticised for want of

dash, and the criticism may be not unfounded: but his grip, when fastened, was that of a vice: and although Britain has had greater generals, to ourselves of the King's Royal Rifle Corps the recollection of this memorable campaign, in which all our battalions under the immediate command of our Colonel-in-Chief took part, must be doubly gratifying from the knowledge that the results of Amherst's operations were more momentous to the British Empire than those even of Wellington or Marlborough.

But on the struggle for mastery the French may look back with no less pride than ourselves. Despite neglect, nay, desertion from home; despite a contest against far superior numbers of population if not of soldiers, no defence of a country has ever been more gallantly conducted. Well may France enshrine the memory of Montcalm, Lévis, and Bougainville. Her sceptre in America has passed away, but the names of her other great sons, Cartier, Frontenac, La Salle, Champlain, La Gallissonnière, &c., are indelibly graven upon the archives of the Western Continent.

The vast territory of Canada had passed into the hands of the British; and, without detracting one iota from the splendid services of other corps engaged, we may feel pride in the thought that from force of circumstances and the number of its battalions, none had a larger share in this conquest—the greatest ever achieved by British arms—than the Royal American Regiment.

CHAPTER V

CAPTURE OF MARTINIQUE, ETC.—AND OF CUBA—END OF THE SEVEN YEARS' WAR

THE British had not long been masters of Canada, when a part of Amherst's troops was required to make further conquests; this time in the West Indies. In January 1759 a first expedition had been sent against the French Antilles, which, though it had failed to master Martinique, had, through the skill and energy of General Barrington, succeeded in capturing Guadeloupe (May 1). In January 1761 Pitt wrote to Amherst requiring 2000 of his men for an attack upon Dominica and St. Lucia; and 6000 more at a later date for a second attempt upon Martinique. The first division of these under Lord Rollo arrived at Guadeloupe in June; and on the 6th of the month captured and occupied Dominica with little trouble or loss. But there the operations for the present ceased; for Martinique was a far more formidable place to assail than Dominica, and other troops were required besides those in America. Barbados was as usual the place of rendezvous; and in the last days of November and the first of December Admiral Rodney's fleet began to arrive by single ships in Carlisle Bay, the port of the island. A few days later arrived the first contingent from Europe; and on Christmas Eve that from America, some 6000 strong under command of General Monckton, also reached the rendezvous. Eleven different regiments combined to make up this number, and among them was the 3rd Battalion of the Royal Americans under Augustin Prevost, by this time Lieutenant-Colonel. Its Colonel-Commandant, Haviland, was in charge of a brigade.

Some days were spent in refreshing the troops and taking in stores; but on January 5, 1762, the armament sailed away to leeward. On the 7th it anchored in St. Anne's Bay on the western side and close to the southern extremity of Martinique. The military force on board now numbered twenty battalions, but seems to have hardly exceeded 8000 men, so deadly had the climate of the West Indies proved even to troops which had known it only for a few weeks. The French, having received intelligence of the expedition, had in the meanwhile spared no efforts to prepare for defence. On their side they could show a garrison of 1200 regulars, 7000 men of the local militia, and 4000 hired privateersmen—no contemptible force in an island of lofty, rugged, volcanic mountains, densely covered for the most part with forest.

As soon as the British fleet had cast anchor, two brigades were landed in the Anse d'Arlet, a bay farther up the western coast, with the object of taking in rear the batteries which defended the entrance to the bay of Fort Royal on its southern side. The road being, however, found to be impracticable for cannon, the brigades were re-embarked; and after sundry feints made by the fleet at different parts of the coast, which exhausted the French by incessant marching, and after heavy cannonading of the shore batteries, the entire force was landed without loss at Case Navire, a little to the north of Negro Point, and as the crow flies hardly more than three miles west of the capital, Fort Royal.

The road, however, passed through deep gullies and

ravines, and the French had erected redoubts on every point of vantage, in addition to batteries upon a hill immediately west of Fort Royal, named Morne Tortenson. Monckton was therefore obliged to throw up batteries of his own to confront them, a task which kept him employed until the 24th. At dawn of that day the British guns opened a heavy fire, under cover of which a general attack was delivered upon the French works. At the same time one brigade was directed to follow the road with a view to taking the coast batteries in rear, and a party of light infantry made its way through the woods to turn the enemy's right flank. The British troops were, for the most part, tried veterans, whereas the French were of the poorest quality. The defence was therefore feeble, and the attack so impetuous that the British carried post after post with little loss. By nine o'clock they were in possession of Morne Tortenson with the whole of its guns and entrenchments. Two brigades simultaneously attacked other of the enemy's posts to the north of Morne Tortenson and, although with great difficulty owing to the steepness of the ground, succeeded in carrying them. Monckton, being now within range of the citadel of Fort Royal, began to construct his batteries; but finding himself harassed by the French guns upon Morne Grenier, a hill which commanded Morne Tortenson, he decided that these must first be taken. The French, however, in a fit of boldness, on the afternoon of the 27th suddenly attacked his left, only to be repulsed with great loss and hunted back to their vantage ground. The British were so keen in pursuit that they followed the fugitives headlong over the steepest and most impracticable country; never pausing, even after night had fallen, till they had captured every work on Morne Grenier, and cleared every

Frenchman therefrom. The batteries on Morne Tortenson were then completed; and after a short cannonade, Fort Royal, which was not yet the strong fortress that it was to become by 1793, was battered into surrender. The whole of the operations had cost the British about 40 officers and 500 men killed and wounded. Since no accounts give the composition of the brigades it is impossible to say what part the Royal Americans took in these affairs. On the 24th they lost 6 men killed and 13 wounded, including Captain William Baillie; on the 27th they lost 6 rank and file killed, 1 ensign and 32 men wounded.

After the fall of Martinique, Monckton, without delay, sent detachments to capture the islands of St. Lucia, St. Vincent, and Grenada, all of which surrendered in the course of February and March. He was meditating further conquests when he received news from home that war had been declared against Spain. Pitt had advocated such a war in 1761, immediately upon learning of the secret alliance known as the Family Compact, which had been concluded between the two branches of the House of Bourbon at Madrid and Paris. His colleagues, however, refused to support him, and he had accordingly resigned office; but his successor, Lord Bute, in spite of his opposition to Pitt's policy, found himself unable to escape from following it, and open hostility with Spain was proclaimed in January 1762. Thereupon Bute decided forthwith to send an expedition under Lieut.-General Lord Albemarle, for the capture of Havana, the capital of the island of Cuba; and Monckton was instructed to detach a portion of his force for the same service. He accordingly sent fifteen of his battalions to join Albemarle, and among them was the 3rd Battalion of the Royal Americans, numbering about six hundred of all ranks, and still commanded by Lieut.-Colonel Augustin Prevost. The battalion, together with four companies of the 42nd, and nine of the 56th, ultimately formed the First Brigade under Brigadier-General Haviland of the 60th.

On April 25 Albemarle arrived at Fort Royal, Martinique, to pick up Monckton's men ; and with a force thus raised to about twelve thousand of all ranks, sailed under escort of a powerful fleet, commanded by Sir George Pocock, for Cuba. Off St. Nicholas Mole in the island of Haiti, the armament was joined by a second squadron which had been cruising to prevent a junction of the French fleet with the Spanish ships at Havana. Finally on May 27 Albemarle weighed anchor and made for Cuba by way of the Bahama Channel. On June 6 Havana was sighted, and at daybreak on the 7th the army landed without opposition at a spot a few miles to the north of the city. On the 8th it advanced westward, brushing aside a force of militia that made a show of resistance, and on the same day arrived before the principal defences of the capital.

The entrance to the harbour lay through a channel about two hundred yards wide, defended at its mouth by two forts—Moro on the northern shore, and Puntal opposite ; the city of Havana itself being situated a little to the east of Fort Puntal. On the north side the ground rises rather abruptly into a ridge called the Cavannõs, at the end of which stands Fort Moro, abutting on the open sea. A detached redoubt on these heights was captured with little loss ; and, since Fort Moro was found to be surrounded by impenetrable jungle, the construction of batteries was begun under cover of the trees, the nearest of them within less than 200 yards from the fort. The difficulty of construction was great, for so rocky was the ground that earth enough to keep the fascines in position could hardly be found. The hardships suffered by the troops in the course of the siege were also very great. No water was to be found near the trenches, and it was only with great difficulty that a scanty supply could be brought from the ships; yet the work to be done was particularly severe, owing to the need of cutting roads through the jungle and the enormous difficulty of dragging guns over a rough, rocky shore. Many men dropped dead with thirst and fatigue, while fever carried off many others, and disabled nearly half the army. The sea was covered with floating corpses.

At length on July 1 the British batteries opened fire from forty-seven heavy pieces on the land side, while four large men-of-war cannonaded the fort from the sea. The ships, after six hours' firing, retired very severely damaged, and for a time the guns of Moro answered Albemarle's batteries with great spirit and effect. By the evening all but two were, however, silenced; but on the next day the principal battery on the British side accidentally caught fire and was wholly consumed. Worse than this, the forest that had masked the batteries was of course by this time cut down, and consequently, during the construction of new works there was no longer shelter from the Spanish fire. New batteries were, none the less, with great exertion constructed, only to catch fire and be again destroyed on the 11th. Yet more batteries were thrown up, and at length on the 19th a lodgment was effected in the glacis under cover of gabions stuffed with cotton, which had been purchased from the homeward-bound fleet from the West Indian Islands. Then a new difficulty presented itself, for the ditch of Fort Moro was found to vary from forty-five to sixty feet in depth, and from forty to one hundred feet in breadth. The British miners, however, under cover of a thin ridge of rock, which had been left to keep the sea from beating into the ditch, made their way to the foot of the wall and began to sink a shaft for a mine by which to blow the counterscarp into the ditch. On the 22nd the Spaniards made a sortie, which was repulsed with heavy loss to themselves and trifling injury to the British.

We are told by Patrick Murray (*vide* p. 294) that the repulse was effected by the 60th, which distinguished itself by the defeat of the 'Dragons de Edinburgo,' and a number of mounted militia. The Royal Americans, having penetrated a morass, charged and defeated the troopers. At the present day the 'Dragons de Edinburgo' are known as the 'Lanceros de Hernan Cortes, 29° de Caballeria.'

On the 29th the mines were reported to be ready. On the 30th at two o'clock in the afternoon they were sprung; and though the breach was narrow and difficult it was discovered by Captain-Lieutenant Fuser of the 60th, A.D.C., and Fort Moro carried by storm with the loss of little more than fifty killed and wounded.[1] Albemarle then pushed forward his preparations against the town; and upon the fall of Fort Puntal on August 10, the garrison agreed to surrender, receiving the honours of war. Our losses during the siege numbered nearly 1800 killed in action, dead of disease, wounded, and taken; and two-thirds of these casualties were due to sickness.

The Royal Americans suffered less from disease than other regiments, losing only 2 officers—one of whom was Lieutenant John McDougal—and 13 men from this cause. From the enemy's fire they had Ensign Stewart and 26 men killed, Lieutenant Sears, Ensign Power

[1] The Spanish casualties were heavy: 130 were killed, 213 others drowned or killed in the boats; 16 officers and 310 men were captured.

and 63 men wounded, and 11 were returned as missing.

One hundred and two brass and two hundred and forty-nine iron guns were taken. The prize exceeded £368,000, of which the Commander-in-Chief's share amounted to £86,030 17*s*. 2*d*., while that of the private soldier was only £2 17*s*. 11*d*.

Pestilence had raged furiously during the siege, and it continued to rage as furiously after the capitulation. On October 18 a return was made of the total losses since the disembarkation in Cuba, when it appeared that over 4700 men had actually perished of sickness alone. Troops sent back to America to regain health suffered almost as severely as those left in garrison. But Cuba was in those days considered the most valuable possession in the world, and it was necessary to keep at Havana a garrison of 6000 men which, it was reckoned, with deduction of one-third in hospital, would leave 4000 efficient.[1] Fortunately a general peace was concluded in the autumn of 1762; Havana was given up, and by July 6, 1763, the last British soldier had left the island.

The Seven Years' War was at an end. After a struggle no less gallant than that which ended in 1815

[1] The following quaint extract, dated 'Head Quarters, 25 August 1762,' from the Orderly Book of Lieut.-Colonel Israel Putnam, at that time commanding a battalion of Provincials and afterwards a General Officer in the War of Independence, shows that care was taken to prevent irregularities and inculcate good manners. It is to be hoped that the grammar, style and spelling are those of Putnam's clerk. and not of the Commander-in-Chief. 'Lord Albemar Recommends it very strongly to the Officers to Prevent their soldiers Bying their Provisions, Liquores Sold by the Inhabitants of this town and to be very attentive to their men's messing Least they should dispose of their Provisions to the town people. Great Decency is to be Shone by the Soldiers to all Religious Professors by Stoping in the Street and Pulling of their Hats Regarly and Soberly will Recommend the Soldiers to the Earl of Albemar who would be sorry they Should Do anything to Selly the good appearance He Has of them.'

France was beaten to her knees. She had lost her possessions in the Far East, as well as in the Far West, and in this very year died Dupleix, the fallen hero of French Hindostan. Choiseul, the great French Minister, knew well that the financial resources of his country were exhausted and that peace was essential. Yet he still hoped that Canada might be given back: and afterwards uttered the prophetic warning that its retention by Great Britain would be followed by American Independence. Many persons in England shared Choiseul's views, and for a time the fate of Canada hung in the balance; but by the terms of the Treaty concluded at Paris on February 10, 1763, France surrendered to Britain the whole of her possessions in North America with the exception of New Orleans and the neighbouring territory.

'George II.,' observed Edmund Burke, a few years later, 'carried the glory, the power, the commerce of England to a height unknown even to this renowned nation in the times of its greatest prosperity: and he left his succession resting on the true and only foundation of all national and all regal greatness; affection at home, reputation abroad, trust in allies, terror in rival Nations. The most ardent lover of his country cannot wish for Great Britain a happier fate than to continue as she was then left.'

When Louis XV. heard of the capitulation of Montreal he remarked, 'What matters the loss of a few acres of snow?' Although in England the acquisition was better appreciated, it is only now, after the lapse of nearly a century and a half, that we are beginning to realise what the cession of Canada meant to the history of our Empire.

CHAPTER VI

OCCUPATION OF THE GREAT LAKES—INDIAN DISCONTENT—PONTIAC RAISES THE STANDARD OF REVOLT—THE SIEGE OF DETROIT—LOSS OF THE OUTLYING POSTS—DEFENCE OF FORTS PITT, LIGONIER, AND BEDFORD

MONTREAL had fallen. Canada, which at that time embraced the whole shores of the Great Lakes and the head waters of the Illinois and Wabash Rivers, was a part of the British Empire. In accordance with the terms of the capitulation it only remained to take possession of those outlying forts and posts, from Ontario to Michigan, where the lilies of France still floated on the breeze.[1]

To this end Major Robert Rogers, an adventurer who in the late war had distinguished himself by daring exploits and hairbreadth escapes, was despatched on September 12, 1760, with 200 men of the 1st Battalion of the Royal Americans. Rogers embarked on the St. Lawrence in fifteen whale boats; and leaving Montreal, proceeded up-stream to Niagara where he quitted his party for the purpose of conveying despatches to General Monckton at Fort Pitt, but rejoined it at Presqu'île. Coasting along the southern shore of Lake Erie, Rogers, on November 7, reached a river which he terms the

[1] The French still retained their settlements on the Mississippi and Wabash Rivers.

Chogage, where through stress of weather he halted and formed an encampment.

It is hardly necessary to say that, at the period of which we are speaking, these outlying regions were occupied by the Red Indian tribes, whose origin is even yet unknown, but who are believed by many to have migrated at some far-distant epoch from Asia. Foremost among them were the Hodenosaunee, better known under their French appellation of Iroquois. The Iroquois comprised the Mohawks, the Oneidas, the Onondagas, the Cayugas, and the Senecas,[1] and were in consequence usually spoken of as 'The Five Nations,' and often retained the name after a sixth, the Tuscaroras, had in 1715 been added thereto. The Iroquois occupied the tract between Lake Ontario and the eastern confluences of the Susquehanna River. Among their confrères they excelled not less in agriculture than in warfare.

Far larger in numbers and in extent of country occupied was the Algonquin family which, surrounding the Iroquois, extended from Hudson's Bay on the north to the Carolinas on the south; from the Atlantic on the east to Lake Winnipeg and the Mississippi on the west. The Algonquin tribes included :—the Delawares who claimed to be the aboriginal inhabitants of the continent and occupied the right bank of the Ohio ; the Shawanoes ; the Miamis and Illinois farther westward ; the Wyandots on the shores of Lake Erie ; the Hurons ; the Ojibwas—sometimes called Chippewas—on the west bank of Lake Huron ; the Ottawas and the Pottawattamies on the eastern bank of Lake Michigan. On the western shore of that lake roamed the Kickapoos, Winnebagoes, and Menominies.

[1] Each of these was known by many other names.

As a whole the Algonquins, whose friendship had been sedulously cultivated by the French, were far inferior to the Iroquois, the traditional friends of England; but they produced, in the Ottawa tribe, a man whose characteristics as a leader made him a pre-eminent factor in the situation. The name of this chief was Pontiac,[1] one whose intellectual capacity, breadth of view, and power of command marked him out as the prime mover in the events about to take place. Pontiac seems to have been on particularly intimate terms with the French commanders, and was present at Braddock's defeat. Although quick to realise that Great Britain must in the future be the dominating power, the very fact filled him with hatred and dismay.

It was from Pontiac that Major Rogers, immediately after pitching his camp, received a visit. Its object was to ask how the Englishman dared to trespass on Indian territory. Rogers replied that the power of France was now superseded by that of Britain; that he was *en route* to occupy Detroit[2] and the other French posts; and that he was about to establish a general peace between white and red men alike. Pontiac made a friendly and polite reply; and after a few days of rest, Rogers resumed his journey. Lieutenant Brehm, of the Royal Americans was now sent forward with a message to Captain Belêtre, French Commandant of Detroit, an important fort situated on the western bank of the strait dividing Lake St. Clair from Lake Erie. Brehm informed the Commandant of the capitulation of Montreal. Rogers shortly afterwards arrived in person and, on November 29, in the presence of 700 Indian warriors, the Fleur de Lys gave place to the Cross of St. George. The Canadian

[1] Sometimes spelt Pondiac or Pontiach.

[2] At that time usually called 'Le Detroit.'

Militia was then disarmed; the Indians standing by in astonishment, partly at the fact of so large a force surrendering to so small a detachment; partly at the, to them, quixotic forbearance of the British in refraining from murdering the Canadians as soon as they had become defenceless.

In pursuance of his mission, Rogers then went on towards Michilimackinac; but winter was already upon him; and though possession was taken by another officer of forts Miamis and Ouatanon, between Lake Erie and the Ohio, Rogers himself was compelled to return to Detroit without achieving his object; and it was not till the following spring that Michilimackinac, with the posts of Sault Ste. Marie, Green Bay, and St. Joseph, passed into British hands.

The task of forming garrisons for these outlying forts was assigned almost exclusively to the 1st Battalion of the Royal Americans, which was thus distributed:—Headquarters and five companies at Lancaster, the remaining five companies detached as follows:

Fort Bedford .	On the eastern slopes of the Alleghany Mountains.	Captain Louis Ourry.
Fort Ligonier .	On the road from Bedford to Fort Pitt.	Ensign A. Blane.
Fort Pitt .	At the junction of the Alleghany and Monongahela Rivers.	Captain Simeon Ecuyer.
Fort Venango .	Some sixty miles due north of Fort Pitt, at the junction of two confluences of the Alleghany River.	Lieutenant Francis Gordon.

Fort Niagara & Schlosser .	At the north and south points of the strait separating Lakes Ontario and Erie, commanding the great entrance to all the interior country.	Major W. Walters (subsequently Major J. Wilkins).
Fort Presqu'île.	On the eastern shore of Lake Erie.	Ensign John Christie.
Fort Le Boeuf.	At end of portage joining French Creek to the Alleghany.	Ensign G. Price.
Fort Sandusky.	At the southernmost point of Lake Erie.	Ensign Christopher Pauli.
Fort Miamis .	At the head of the Maumee River.	Ensign R. Holmes.
Fort Ouatanon.	A hundred miles to the south - west, on the Wabash, giving command of the great trade route from Lake Erie to the Ohio.	Lieutenant E. Jenkins.
Fort Detroit [1] .	On the western bank of the strait separating Lake Erie from Lake St. Clair.	Captain Donald Campbell and Lieutenants J. Hay, Macdonald and G. McDougal.
Fort Michilimackinac (sometimes abbreviated to Mackinaw).	Which secured free communication between Lakes Huron and Michigan.	Captain G. Etherington, Lieutenants W. Leslie and Jamet.
Fort La Haye &	At the southernmost point of Green Bay.	Ensign J. Gorrell.
Fort St. Joseph	On the south-eastern shore of Lake Michigan.	Ensign F. Schlösser.

(The last two guarded the respective routes to the Mississippi by the Wisconsin and Illinois Rivers.)

[1] Major H. Gladwyn, who had been for a few months in the 60th, but was now in the 80th Light Infantry, commanded the garrison, which comprised a company of his own regiment as well as one of the Royal Americans.

The vast extent of country to be guarded formed a striking contrast to the slender force appointed for the purpose. Some of these detachments consisted of a mere handful of men. Captain Ourry, on one occasion, reported the desertion of 'the sixth part of my garrison,' a phrase stronger in appearance than in reality; for the 'sixth part' turned out to be but one man! To a person of foresight and experience the situation might well appear to be fraught with danger; but having broken the power of the French it does not seem to have occurred to Sir Jeffery Amherst that any peril was to be feared from the original occupants of the soil, crafty and treacherous though they were known to be. Yet warnings of discontent seething beneath the surface were not wanting. As early as June 17, 1761, Captain Donald Campbell, in temporary command at Detroit, evidently appreciating the gravity of his news, wrote to Major Walters at Niagara to say that he had been alarmed by reports of bad designs on the part of the Indians, and was now in possession of reliable information to the effect that it was the intention of the Senecas to assemble at French Creek before the end of the month, and surprise Niagara; while the Delawares and others had arranged simultaneously to surprise Fort Pitt and sever the communications between the outlying posts. Not content with this report Campbell further despatched messengers to the commanders of posts, as well as to Amherst; and the plot having been discovered, was for the moment nipped in the bud, as was another formed in the following year.

But Amherst was deaf to all warning or remonstrance. He would neither take steps to conciliate the tribes nor decide on military precautions. Preparations were made for the disbandment of our 3rd and 4th Battalions, while the 1st and 2nd were each reduced to nine companies and a joint strength of 1000 men.

Wearily and drearily passed the days for the occupants of the outlying forts, although at times beguiled by the charms of Indian beauties; at others by fishing, hunting, and marksmanship. Provisions, too, ran short. The stores at Fort Miamis in the spring of 1762 were reduced to 10 lb. of powder, 2 barrels of flour, 1½ barrels of pork, 7 or 8 bushels of Indian corn. The Indians would bring no meat, and Ensign Holmes had no resources but to make a contract with a sutler.

'We are so miserable here,' writes Ensign Schlosser from St. Joseph to Colonel Bouquet, on January 24, 1762, 'that I have never in all my life seen a soldier actually in service suffer so much by want as we suffer without distinction. We have no kind of flesh nor venison nor fish, nothing to hound (*sic*); and that we could suffer with patience, but the porck is so bad that neither officer nor men can eat it, and self lief [I myself have lived] more than seventeen weeks up[on] flour and peace soup, and have eat no kint of meat but a little bear at Christmas. We have plenty to drink, and that I think is what kips up in health, and the bread which is tolerably good.'

Should the hypercritical reader comment on the orthography of the German, it is only fair to say that the spelling of the Scottish officers, of whom there were many in the Regiment, also left something to be desired. One of them, writing on the same subject from Fort Ligonier, complains that 'a great deal of flour has been destroyed by the Katts which,' he adds, 'are enumerable.'

The unfriendly attitude of the Indian tribes was accentuated by the misguided spirit of economy displayed by Amherst in stopping the practice of giving presents to the chiefs. In vain did Colonel Bouquet and Sir William Johnson, than whom no one in America was better acquainted with the temperament of the Indians, point out the probable consequences of such parsimony. Sir Jeffery refused absolutely to listen; and the neglect and contempt which he showed for the Red-skins was, it is to be feared, shared by the British soldier, whose contemptuous treatment of the Indians formed an unpleasing contrast to the tact and flattery which had been so successful an instrument in the hands of the French. The men of the Royal Americans who recalled the massacre of Fort William Henry, who fought both beside and against Indians at Ticonderoga and Quebec, had now for the most part been discharged, and their places taken by raw recruits; so that in lack of experience, no less than in numbers, the Regiment as a whole was not quite what it had been.

What Amherst failed entirely to realise was the fact that the Iroquois and Algonquins, although ignored in the Capitulation of Montreal, considered themselves still the masters of the soil. They readily owned that they had ceded a portion to the French, but refused to acquiesce in the view that the French could give any title to another occupant; and their resentment at the quiet way in which the British, without saying so much as 'by your leave,' took possession of the country and covered it with settlements, was great. The irritation was increased by the Canadians who, although they had now sworn allegiance to the British Crown and had consequently been allowed to retain their farms, left no stone unturned to increase the rising animosity by stories that the French Monarch had been asleep, but being now awakened, was sending irresistible forces to recover his lately lost dominions. Lieutenant Jenkins, writing from Ouatanon in March 1763, to the officer commanding at Detroit, says that the Canadians there were eternally telling lies to the Indians ; declaring that a great army would come from the Mississippi, and capture first the outlying posts, and afterwards regain Canada. Holmes, at Miamis, having received a warning, took the bold course of assembling the Indian warriors and charging them with their design. They professed contrition, and a crisis was for the moment averted.

But shortly afterwards came the news that at the Peace of Paris the French Monarch, without any reference at all to the Indians, had ceded to the British the whole of the French territory east of the Mississippi. It was a terrible shock, and brought matters to a crisis. Even at the eleventh hour tact and diplomacy might have brought about a satisfactory result, for the British Government had given orders that the valley of the Ohio was to be occupied by none but Indians ; but Amherst was still deaf to all remonstrance. Headed by Pontiac, the Ottawa chief already noticed, who was burning with indignation and vowing vengeance at the slights which he had received from our Government, the Algonquin tribes, aided by the Wyandots and the Senecas (one of the Six Nations hitherto loyal to the British), prepared to rise in revolt.

Between the circumstances attending this outbreak and those of the Sepoy mutiny of 1857, a parallel can be closely drawn. In both cases was the conspiracy organised with amazing skill. In both cases was all warning unheeded, over-confidence was rampant, the available force was a mere handful; and in both cases was the temporary triumph of the insurgents marked by ghastly scenes of butchery. In both cases, too, the 1st Battalion of the 60th played a leading part.

The number of British troops available on the Great Lakes was ridiculously inadequate, consisting as it did of one or two companies of the 80th L.I., a few men of a Provincial Corps termed the Queen's Rangers, and a weak wing of our 1st Battalion. The 2nd Battalion was stationed on the bank of the St. Lawrence. The 3rd and 4th were in process of disbandment; so was the 80th Regiment. The other wing of the 1st Battalion of the 60th, which together with one or two skeleton regiments formed the sole reserve, was 500 miles distant at Lancaster.

On April 27, 1763, Pontiac, at a general meeting of the Indian tribes, sounded the war-cry. With fierce gesture and in impassioned tones he recited the grievances of his race; contrasted the conduct of the British with that of the French; described the insulting manner in which the Indians had been treated by the soldiers of the present garrisons; and declared that the destruction of the French power would be followed by the extermination of the Indians.

> 'Then, holding out a broad belt of wampum,[1] he told the council that he had received it from their great father the King of France, in token that he had heard the voice of his red children: that his sleep was at an end: and that his great war canoes would soon sail up the St. Lawrence to win back Canada, and wreak vengeance on his enemies. The Indians and their French brethren would fight once more side by side, as they had always fought; they would strike the English as they had struck them many moons ago, when their great army marched down the Monongahela, and they had shot them from their ambush like a flock of pigeons in the woods.'

Pontiac's speech was received with acclamation, and immediate action was demanded.

[1] The name 'wampum' was given to belts or necklaces made of broken bits of shell strung together. When a belt of wampum was circulated as a summons to war, its colour was always red or black. When used for other purposes, the shells were white.

It was at Detroit that the storm burst. There the fort was built in the form of a parallelogram, whose four sides were 880 yards in all. The defences consisted of a picket fence twenty-five feet high, with a blockhouse over the gateway. Two small armed schooners lay in the stream. Settlers in the neighbourhood, on both sides of Lake St. Clair, numbered about a thousand; but the actual garrison consisted of a company of the 80th and Phillips' company of our 1st Battalion. Whether Phillips himself was present seems doubtful. Captain Donald Campbell was there, and had as his subalterns Lieutenants MacDougal and James MacDonald.

The following letter, docketed by Colonel Bouquet, 'Lettre du Lieutenant Macdonald, datée au Detroit le 12e juillet 1763,' describes the sequel in graphic terms, even if the spelling be open to criticism.

Detroit. 12 July 1763.

'Sir,

'You have certainly heard long before now of our misfortunes at the Detroit and its depentancies, but as it may be satisfactory for you to be more particularly informed, will do myself the pleasure to give you an exact account of all that has happened in this department, and hope you'll do me the justice that I would have wrote you, and communicated the same long ago had an opportunity offered.

'On Friday the 6th of May we were privately informed of a conspiracy formed against us by the Indians, particularly the Ottawa nation, who were to come to council with us the next day, and massacre every sould of us. On the morning of that day, being Saturday the 7th of May, fifteen of their warriors came into the fort and seemed very inquisitive and anxious to know where all the English merchants' shops were. At 9 o'clock the garrison were ordered under arms, and the savages continued coming into the fort till 11 o'clock, deminishing their numbers as much as possible by dividing themselves at all the corners of the streets most adjacent to the shops. Before 12 o'clock they were three hundred men, at least three times in number equal to that of the garrison. But, seeing all the troops under arms and the merchants' shops shut, I imagined prevented them from attempting to put their evel scheme in execution that day. Observing us thus prepared, their chiefs came in a very condemned-like manner to council where they spoked a great deal of nonsense to Major Gladwin and Captain Campbell, protesting at the same time the greatest friendship imaginable to them, but expressing their serprise at seeing all the officers and men under arms. The Major then told them that he had certain intilligence that some Indians were projecting mischief, and on that account he was determined to have the troops always under arms upon such occasions; that they, being the oldest nation and the first that had come to council need not be astonished at that precaution, as he was resolved to do the same to all nations. At 2 o'clock they had done speaking, went off seemingly very discontented, and crossed the river half a league from the fort, where they all encamped. About 6 o'clock that afternoon six of their warriors returned and brought an old squaw prisoner, alledging that she had given us false information against them. The Major declared she had never given us any kind of advice. They then insisted upon naming the author of what he had heard with regard to the Indians, which he declined to do, but told them that it was one of themselves whose name he promised never to reveal, whereupon they went off and carried the old woman prisoner with them. When they arrived at their camp, Pondiac, their greatest chief seized on the prisoner and gave her three stroks with a stick on the head, which laid her flat on the ground, and whole nation assembled round her and called repeated times, Kill her, kill her.[1]

'Sunday the 8th, Pondiac and several others of their principle chiefs came into the fort at 5 o'clock in the afternoon, and brought a pipe of peace with them, with which they wanted to convince us fully of their friendship and sincerity, but the Major, judging that they only wanted to caggole us would not go neigh them nor give them any countenance, which obliged Captain Campbell to go and speak with them; and after smoking with the pipe of peace and assuring him of their fidelity, they said that the next morning all the nations would come to council, where everything would be settled to our satisfaction, after which they would immediately desperse, and that would remove all kind of suspicion. Accordingly,

[1] The lady survived this severity many years, and might have lived still longer but for the misfortune of falling into a butt of boiling maple-sap when in a state of intoxication.

on Monday morning, the 9th, six of their warriors came into the fort, at 7 o'clock, and upon seeing the garrison under arms, went off without being observed. About 10 o'clock we counted fifty-six canoes, with seven or eight men in each, crossing the river from their camp, and when they arrived neigh the fort the gates were shut, and the interperter sent to tell them that not about fifty or sixty chiefs would be admitted into the fort, upon which Pondiac immediately desired the interperter in a prepyrtor manner to return directly and acquaint us that, if all their people had not free access into the fort, none of them would enter it; that we might stay in our fort but he would keep the country, adding that he would order a party instantly to an island where we had twenty-four bullocks, which they immediately killed. Unluckilly, three soldiers were on the island, and a poor man with his wife and four children, which they all murthered, except two children, as also a poor woman and her two sons that lived about half a mile from the fort. After having thus put all the English without the fort to death, they ordered a Frenchman who had seen the woman and her two sons killed and scalped, to come and inform us of it. . . . We were then fully perswaded that the information given us was well founded, and a proper disposition was made for the defence of the fort, although our number was but small, not exceeding one hundred and twenty, encluding all the English traders and the works very neigh a mile in circumference.

'On Tuesday the 10th, early in the morning, the savages began to fire at the fort and vessels which lay opposite to it. About 8 o'clock the Indians called a parley and ceased firing, and half an hour after, the chief of the Wiandotes came into the fort on their way to a council where they were called by the Ottawas, and promised us to endeavour to solicite and perswade the Ottawas from committing further hostilitys. After drinking a glass of rum they went off.

'At 3 o'clock that afternoon, severals of the inhabitants and four chiefs of the Ottawas, Wiandotes, Chippawas and Pattawattamies came and acquainted us that most of the inhabitants were assembled at a French mano-house about a mile from the fort, where the savages purpose to hold a council, and desiring Captain Campbell and other officers to go with them to that council, where they hope, with their presence and assistance, further hostilities would cease; assuring us, at the same time, that be as it would, that Captain Campbell and the other officer that went with him should return when ever they pleased. This promise was assertained by the

French as well as the Indian chiefs. Whereupon Captain Campbell and Lieut. McDougall went off, escorted by a number of the inhabitants and the four chiefs (the first promised to be answerable for their returning that night). When they arrived at the house above mentioned, they found the French and Indians assembled, and after counselling a long time, the Wiandotes were prevailed on to sing the war song, and this being done, it was next resolved that Captain Campbell and Lieut. McDougall should be detained prisoners, but would be indulged to lodge in a French man's house till a French commandant arrived from the Illinoise; that next day five Indians and as many Canadiens would be dispatched to acquaint the commanding officer at the Ilinoise that Detroit was in their possession, and require of him to send an officer to command, to whom Captain Campbell and Lieut. McDougall should be delivered. As for Major Gladwin, he was summoned to give up the fort and two vessels, etc., the troops to ground their arms; that they would allow as many battoes and as much provisions as they judged requisite for us to go to Niagara; that, if these proposals were not excepted of, they were a thousand men and would storm the fort at all events, and in that case every soul of us should be put to the torture. The Major returned for answer that, as soon as the two officers they had detained, were permitted to come into the fort, he would, after consulting them, give a positive answer to their demands, but could do nothing without obtaining their opinion. On Wednesday, the 11th, several inhabitants came early in the morning into the fort, and advised us, by way of friendship, to make our escape aboard the vessels, assuring us we had no other method by which we could preserve our lieves as the Indians were then fifteen hundred fighting men, and would be as many more in a few days; and that they were fully determined to attack us in an hour's time. We told the Mons[es] that we were ready to receive them and that every officer and soldier in the fort would willingly perish in the defence of it, rather than condecend, or agree to any terms that savages would propose; upon which the French went off, as I suppose, to communicate what he had said to their allies, and in a little afterwardes the Indians gave their usual hoop, and about five or six hundred began to attack the fort on all quarters. Indeed some of them behaved extremely well, advancing very boldly in an open plain, exposed to all our fire, and came within sixty yards of the fort; but, upon having three men killed and about a dozen wounded, they retired as briskly as they advanced, and fired at three hundred yards distance till 7 o'clock at night, when they sent a Frenchman into the fort with a letter for the Major desiring a cessation of arms that night, and proposing to let the troops, with their arms, go on board the vessels, but insisted upon our giving up the fort, leaving the French artillery, all the merchandize and officers' effects, and even the insolence to demand a negro boy belonging to a merchant to be delivered to Pondiac. The Major's reply to these extraordinary proposition was much the same as to the first.

'Thursday, the 12th, five French men and as many Indians were sent off for the Ilinoise with letters wrote by a Canadien agreeable to Pondiac's desire. On the 13th we were informed by the inhabitants that M. Chapman, a trader from Niagara was taken prisoner by the Wiandotes with five battoes loaded with goods. The 21st, one of the vessels was ordered to sail for Niagara, but to remain till the 6th of June at the mouth of the river, in order to advert some battoes which we expected daily from Niagara. Upon the 22nd we were told that Ensign Paullie, who commanded at Sanduskey, was brought prisoner by ten Ottawas, who reported that they had prevailed, after a long consultation with the Wiandotes, who lived at Sanduskey, to declare war against us; that some days ago they came, early of a morning, to the blockhouse there, and murthered every sold therein, consisting of twenty-seven persons, traders included; that Messrs. Callander and Prentice, formerly captain in the Pensylvannia regiment, were amongst that number; and that they had taken one hundred horses, loaded with Indian goods, which, with the plunder of the garrison, was agreed on to be given the Wiandotes before they condescended to join them; that all they wanted was the commanding officer. . . .

'Sunday, the 5th of June, we were acquainted that Fort Miamis was taken. . . .

'The 10th of June we heard that Ensign Schlosser, the commanding officer at St. Joseph's was taken prisoner, and that all his garrison (except three men) were massacred.

'The 12th we were told that Lieut. Jenkins and all the garrison of Ouiattanon, consisting of a sergeant and eighteen men, were taken prisoners and carried to the Ilinoise.

'The 18th a Jesuite arrived from Michilimackinac and brought us letters from Captain Etherington and Lieut. Lesley, with an account of their being taken prisoners, that Lieut. Jamenat was killed and twenty-one soldiers. . . .

'On the night of the 2nd instant, Captain Campbell and Lieut. McDougall were lodged at the house above mentioned, about a mile from the fort, and made a resolution to escape, when it was agreed and between them that McDougall should set off first, which he did and got safe into the fort. But you know it was much more dangerous for Captain Campbell than for any other person, by reason that he could neither run nor see; and being sensible of that failing, I am sure prevented him from attempting to escape.

'The 4th a detachment was ordered to distroy some breast works and entrenchment the Indians had made, a quarter of a mile from the fort, and about twenty Indians came to attack that party, which they engaged but were drove off in an instant with the loss of one man killed (and two men wounded), which our people scalped and cut to pieces. Half an hour afterwards the savages carried the man they had lost before Captain Campbell, stripped him naked, and directly murthered him in a cruel manner, which indeed gives me pain beyond expression, and I am sure cannot miss but to affect sensibly all his acquaintances. Although he is now out of the question, I must own I never have, nor never shall have a friend or acquaintance that I value more than him. My present comfort is that if charity, benevolence, innocence and integrity is a sufficient dispensation for all mankind, that entitles him to happyness in the world to come.

'Notwithstanding our being attacked almost every day, and often in the night, for two months past, the loss we have sustained is very inconsiderable, being only one man killed and a dozen wounded, whereas the enemy have had about twenty men killed and thirty wounded. I believe they begin to be quite despondant, and are affraid their attempts will prove abortive and to no purpose. I make no doubt but your patience will be entirely exhausted reading so long an epistle, and the more so as the contents thereof is wholly melancholy and disagreeable, and therefore will add no more, but that

'I am, with greatest respect, Sir,

'Your most obedient and faithful servant,

'James MacDonald.

'I had almost forgot to acquaint you that the vessel arrived here the 30th of June from Niagara, and brought us a reinforcement of an officer and fifty men, one hundred and fifty barrils of provission, with plenty of ammunition, which put our garrison in good spirits, I believe the savages begin to see their attempt will prove abortive and to no purpose, and repent their engaging in a war that I hope will end in their total distruction. God grant that may be so. . . .

'Ensign Christy and Schlosser have been brought into the fort by the Pottowattomies and Wiandotes; the first was delivered the 9th instant and the last the 14th of June.

'I shall embrace the first opportunity of sending you a return of our detachment, and would have sent it now, but time will not permitt, as the vessel is just setting sail for Niagara.

'I am respectfully, Sir,

'Your most humble servant,

'J. M. D.'

We must now for the moment leave Detroit and glance at the events taking place at the other forts.

Sandusky

The next to be attacked was Fort Sandusky, some fifty miles distant by water. On May 16, Ensign Pauli, the commandant, who had apparently received no news of what was taking place at Detroit, was told that seven Indians wished to speak to him. They belonged chiefly to the Wyandot tribe; and, as he knew several of them personally, Pauli had no hesitation in admitting the party. After a conversation an Indian standing in the doorway gave a sudden signal. In a moment Pauli was overpowered, disarmed, and bound. At the same time shrieks, yells and the report of guns outside were heard; and the luckless officer being led out into the courtyard found it strewn with the bodies of his murdered soldiers. After seeing his fort burned to the ground, he was carried to the Indian camp at Detroit.

St. Joseph

Further disasters and massacres quickly followed. At Fort St. Joseph, on the southern bank of Lake Michigan,

Ensign Schlosser still led his monotonous life, amid his party of fourteen men. He was not, however entirely destitute of Canadian neighbours, for the place had long been the seat of a Jesuit mission, and the tribe of Pottawattamies, who occupied the neighbourhood, were probably reckoned as Christians. Here, on May 25, the same stratagem was practised as at Sandusky. A warning at the last moment caused Schlosser to call his men to arms and attempt to assemble the Canadians. But it was too late. The Indians within the fort instantly tomahawked the sentry at the gate and let in their friends outside, Eleven of the Royal Americans were murdered. Schlosser and the three remaining privates were bound and taken to Detroit.

Miamis

At Miamis Ensign Holmes had long been on his guard, and a different method was adopted. A young Indian girl solaced his loneliness; and she, on May 27, begged him to come to the aid of a squaw dangerously ill hard-by. Holmes unfortunately believed her story, and on entering the wigwam pointed out to him, was shot dead. His sergeant, hearing the report of the guns, came out to find the cause and was at once captured. Deprived of its leaders the little garrison nevertheless prepared for fight, but on the representation of Canadians that resistance was hopeless, and on promise of their lives the men surrendered.

Ouatanon

Lieutenant Jenkins and nineteen men at Ouatanon were captured by the stratagem which had been fatal to Pauli at Sandusky; but the Miamis apologised for the act, which they alleged they had done under compulsion from their neighbours. The remaining handful of men surrendered and all were well treated.

Michilimackinac

Michilimackinac (known sometimes as Mackinaw—so called from the Indian name for a green turtle, whose back an island in the strait was supposed to resemble) felt the next blow. At this fort were quartered, as already noticed, Captain George Etherington, Lieutenant W. Leslie and Ensign Jamet (or Jamette), with a garrison of some five-and-thirty men. This place again was an old mission station of the Jesuits, and of the two neighbouring tribes, the Ottawas and the Ojibwas (often called the Chippewas), the former were largely (in name) Christians, and in some respects not uncivilised. Both were devoted to the French and consequently hostile to the British.

Of the attack on Detroit the garrison had heard nothing. Etherington is described as a man of great address in dealing with savages, and was probably aware of his popularity. At all events he had more confidence in the Indians than in the Canadians, and turned a deaf ear to the warnings of the latter. The result may be described in his own words:

> 'On the 2nd instant [June] the Chippewas, who live in a plain near this fort, assembled to play ball, as they had done almost every day since their arrival. They played from morning till noon; then, throwing their ball close to the gate, and observing Lieutenant Lesley and me a few paces out of it, they came behind us, seized and carried us into the woods.
>
> 'In the meantime the rest rushed into the fort, where they found their squaws, whom they had previously left there with hatchets hidden under their blankets, which they took, and in an instant killed Lieutenant Jamet and fifteen rank and file, and a trader named Tracy. They wounded two, and took the rest of the garrison prisoners, five of whom they have since killed.'

It so happened, however, that the Ottawas were mightily offended with the Ojibwas for acting without their sanction. The prisoners were, for the moment, rescued; but the quarrel having been made up, a portion were handed over to the Ojibwas. It is possible that their lives might have been spared, but as misfortune had it, a well-known chief appeared on the scene, and, desirous of showing his approval of the recent exploit, murdered seven of the soldiers, who were forthwith eaten by the savages.[1]

But by this time, a certain fear of English retribution began to spread. Partly on account of this, partly through the untiring efforts of Father Jonois, a Jesuit missionary, the thirteen prisoners retained by the Ottawas were well treated. Among them were Etherington and Leslie. The former wrote an order to Lieutenant Gorell at Green Bay to abandon the post and join him. Gorell, with his seventeen men, had been for the last two years in a position of considerable responsibility, and had acquitted himself with credit. He had shown address in conciliating the Indians and in regulating the fur trade among them. As a result his fort was never attacked, and on receipt of Etherington's letter, he left it in charge of the Menominee tribe. On June 21 Gorell and his party embarked on the lake escorted by the Menominees, and on the 30th joined Etherington, Leslie and their eleven men at the village of L'Arbre Croche, some twenty miles south of Michilimackinac. Gorell kept a good look out, and resolutely refused to give up his arms.[2] This fact, coupled with the request of the Menominees, induced the Ottawas to release Etherington's party: and the small combined force, having

[1] To eat a man after killing him was considered by the Indians a compliment to his bravery, which the mastication of his flesh would impart to themselves.

[2] Major Murray gives the credit of this refusal to Sergeant Noel or Null: *vide* p. 290.

on July 18 embarked at L'Arbre Croche, reached Montreal in safety on August 13.

By this time, excepting at Detroit, the British flag had disappeared entirely from the borders of the Great Lakes.

Presqu'île

At Presqu'île Ensign Christie heard the story of massacre and disaster at least as early as June 3. His force was respectable, consisting, as it apparently did, of twenty-four men of his own and six of another regiment. He lost no time in writing to Major Wilkins at Niagara for more provisions and ammunition, and made preparations against attack. The fort was furnished with a blockhouse constructed of heavy logs and believed to be impregnable; but its weak point lay in the fact that it was situated between Lake Erie and a stream, whose steep banks gave cover to the assailants up to within forty yards of the stockade.

At dawn on June 15 two hundred Indians detached from Detroit were seen crossing the mouth of the stream and taking cover under the bank of the lake. Christie, carrying forbearance to an extreme, gave orders that no shot was to be fired except in reply to that of the enemy; and thus at the outset gave the Indians an advantage of which they were not slow to avail themselves. Creeping into the ditch and even into the fort unmolested, they at once opened fire from their shelter upon the loop-holes of the blockhouse at a few yards' distance. Burning arrows and pitch fire balls were hurled on to the roof and woodwork. Again and again the roof caught fire. Again and again the flames were extinguished, but the water barrels became emptied; and the well in the parade ground, the only source of replenishment, was commanded by the enemy's fire. A subterranean passage thereto was however begun, but long before it was complete the roof was once more set on fire, and the fort only saved by the gallantry of a soldier who tore away the burning shingles.

Meanwhile the Indians, working throughout the night and following morning, made a rather scientific attempt to undermine the blockhouse; and with so much success that in the afternoon of the 16th, they reached and set on fire the hut of the commanding officer. The flames spread to the adjoining bastion of the blockhouse, but by this time the underground passage to the well had been finished. Water was procured and the fire once more extinguished. Throughout the day and evening the firing continued without cessation, but at midnight negotiations were opened by the assailants. A message was sent to Christie through the medium of an Englishman who having been captured by the Indians some years previously was now fighting in their ranks, to the effect that preparations had been made to set the blockhouse on fire from two sides at once; that further resistance was useless; that the lives of the garrison would be spared if they surrendered, but that otherwise they would be burned alive. Christie was allowed till morning to consider his answer, and consulted his men. Benjamin Grey and David Smart, two private soldiers of the Royal Americans whose names deserve record, bravely said they would bear the heat as long as they were able and would then fight their way through. Exhausted as the rest were, all but two voted to continue the defence. At dawn two men were despatched to ascertain the truth or falsehood of the Indians' statement. They reported that the assertion was correct and that the blockhouse would be burned down. Christie then surrendered on condition of being allowed to retire unmolested to the nearest British post. The scorched and haggard soldiers issued forth and at once found themselves prisoners in the hands of their savage foe, who had no intention whatever of keeping faith with them.[1] Gray and one other escaped : the rest were taken to Detroit, where several of them seem to have saved their lives by enrolment among the Indian tribes !

At the moment Colonel Bouquet strongly censured Christie for his surrender. He was probably unacquainted with all the facts of the case. A Court of Inquiry was afterwards held on the circumstances of his capitulation ; and it was evidently felt that it would be hard to deprive Christie of his commission. He remained for some years longer in the Regiment, and served therein, not without distinction, during the American War of Independence.

Fort Le Boeuf

The ill-constructed blockhouse somewhat loftily termed Fort Le Boeuf was occupied by Ensign George Price,[2] with two corporals and eleven private soldiers. Part of their powder was damaged, and at best they had only twenty rounds of ammunition per man. On the morning of the 18th a small party of Indians made its appearance from the direction of Presqu'île and asked for ammunition, which was at once refused. Other savages, making in all about five and thirty, then came up and requested admission. This was also refused, and the Indians thereupon made their way into the cellar of a storehouse only ten yards distant, where they were in a position of vantage,

[1] In his subsequent report to the Officer Commanding Fort Pitt, Gray stated that the blockhouse had been undermined and was ready to be blown up.

[2] Price was still an Ensign in 1770, after which year his name disappears from the regimental list.

for the loopholes of the blockhouse were too high to enable the garrison to fire down upon them, while the Indians on the other hand had no difficulty in discharging their arrows upwards. It was not, however, till late in the day that they began to shoot volleys of burning arrows against the roof and sides of the fort. Three times the roof was set alight, and as often extinguished; but at nightfall the blockhouse caught fire once more, and this time the men were unable to put it not. Terrible as was the situation Price remained undaunted. 'We must fight as long as we can and then die together' was his reply to those who urged him to leave his post. But the flames rapidly spread and, when all but suffocated, he gave leave to his men to save themselves if they could. Even then it was with difficulty that he was persuaded to go with them; but at length, taking advantage of a narrow window at the back, through which they safely crept, all escaped from the burning fort. Halting for a moment to take breath they looked back, gazed on the flames; heard the yelling of the Indians and the reports of their firearms, as they continued the attack unconscious of the escape of the garrison. It is possible that the Red-skins might have been taken by surprise in a counter-attack: but the soldiers were no doubt too much exhausted to make any such effort.

Presqu'île was but fifteen miles distant; but from the direction whence the savages had originally come, Price felt sure it had been lost, and determined to make for Venango. His provisions comprised but three biscuits per man. One soldier thought he could guide them but soon lost his way. Six men separated from the rest in the darkness, and when the day broke could not be found. Then Price discovered that after all the toil and wandering he was even now only two miles from Le Boeuf. Forty miles as the crow flies—fifty by the traders' track—separated the party from Venango. Nothing remained to the exhausted men but to push on for dear life. Forward through the heat of the day they bent their weary steps. The sun went down, but through the night they still pushed on. At 1 A.M. on the 20th[1] the party reached Venango, and with feelings better imagined than described found the fort burned to the ground, and amid the embers the half-consumed bodies of their comrades. Fort Pitt, the next fort, was some hundred miles distant; but animated no doubt by the indomitable spirit of its leader, the little party once more pushed forward, and early on the 25th, more dead than alive, arrived at Fort Pitt. Of their six comrades who had been lost in the third night, two died of hunger; the other four were eventually fortunate enough to reach Fort Pitt.[2]

Surely none but men amply endued with the 'hunters' instinct' could have found their way through the unknown forest and morass; and none but men of the utmost hardihood and power of endurance could, despite the pangs of hunger, have reached this goal.

Venango

What had happened at Venango no survivor remained to tell. Colonel Bouquet blamed Lieutenant Gordon, the officer in command; but it seems doubtful whether he had any very precise knowledge of what had occurred: for our information even at the present day is derived chiefly from the story, told many years afterwards, by an

[1] Thus, Parkman. But the 21st seems more probable.

[2] In view of the idea that the 60th was at this period composed of foreigners, it is interesting to note the names of six of Price's party (that of the seventh is unknown), viz. Corporals Jacob, Fisher and John Nash; Privates John Dogood, John Nigley, John Dortinger and Uriah Trunk. The three names of the Presqu'île garrison handed down are, Gray, Smart, and Smyth.

Indian who had been present. His tale, which may be accepted as substantially correct, was to the effect that a large party of Senecas had gained admittance as friends, had put the garrison to death, and reserved Gordon for a more awful fate. The unhappy officer was brought out every evening and partly roasted before a slow fire. It was not until the fifth day that death ended his sufferings.

Fort Pitt

The formidable garrison of Fort Pitt, comprising some 330 men made up of Royal Americans, traders, and backwoodsmen, was commanded by a gallant officer, in whom Colonel Bouquet evidently reposed the utmost reliance, viz. Captain Simeon Ecuyer, a native of Switzerland. Early in May Ecuyer had received warning from Gladwyn of the critical situation, and was well on the alert. On the 27th a party of Indians made its appearance, shot two soldiers near the fort and murdered a good many civilians. More than two hundred women and children took refuge inside the fort, which was so crowded that Ecuyer feared disease, the rather that small-pox had already made its appearance,

The defences had been partly destroyed by floods. The surrounding woods seemed alive with Indians who, having established themselves on the banks of the Ohio and Monongahela Rivers, kept up a desultory though galling fire night and day upon the sentries; but it was not until the afternoon of June 22 that a general attack was made, and repelled chiefly by shell fire from the howitzers of the fort. Next morning a deputation made its appearance at the gate to propose the evacuation of the post. Ecuyer in his reply had recourse to his imagination, informing his audience that three large British armies were already in the field and would very soon exact vengeance on Pontiac and his followers. For his own part he meant to stay where he was. The deputation withdrew, considerably alarmed at the fictitious news, and for the time being, raised the siege. On the 26th Private Gray, who had, as we have seen, escaped from Presqu'île, arrived, bringing news of the many disasters which had taken place,

Until nearly the end of July no further attack was made, and Ecuyer, although closely blockaded and entirely cut off from the outer world, was able to turn his attention to the strengthening of his defences and the security of his women and children. On the 26th another Indian deputation again urged Ecuyer to abandon his post, and received another uncompromising refusal. Much incensed the savages lost no time in making a desperate effort. After dark on the same day they took post under the banks of both rivers : and digging holes therein close to the fort kept up an incessant fire, often with burning arrows. But the garrison, protected by their log parapet, also kept under cover, and during the five days of the attack had a loss of but one man killed and seven wounded. Among these unfortunately was Ecuyer himself, who had been everywhere conspicuous, 'directing, encouraging and applauding the men in his broken English.' Perhaps his most difficult task was to restrain the impetuosity of the garrison, and stop his men from making a sortie in order to come to close quarters with the Indians who were too wily to make a general assault upon the works. In one of his letters (all written in French and with a touch of humour) he tells Colonel Bouquet that he let no one fire until he had marked his man, and felt certain of having killed and wounded twenty of the enemy. Others were probably hit without being noticed.

On August 1 the Indians raised the siege and marched

eastward to meet the avenging Bouquet, who was by this time at hand with a relieving force.

Ligonier

At Fort Ligonier Lieutenant Archibald Blane was in command of a party composed of three men and two boys; by his own account not more formidable in quality than quantity. 'My garrison,' he writes, 'consists of Rodgers unfit for any kind of fatigue; Davis, improper to be trusted on any duty; Shillem, quite a little boy; my servant, the same; and Jones, an inactive simple creature. . . . Were I to send out the five at once two stout fellows would beat the whole of them.' Under these circumstances Blane may not be very severely criticised if, during a moment of weakness, he gave Captain Ourry at Fort Bedford a hint that he might find it desirable to evacuate his fort and join the Captain. But Blane quickly recovered his firmness; he was reinforced and the defence of his fort became memorable. Bouquet wrote to applaud his courage; but, having heard the rumour, added the following words—meet to be engraven on the heart of every rifleman: 'If an officer should remain alone in his post there he must die before he disgraces himself by abandoning it.'

Blane's garrison was augmented by a few pack-horse men, some of whom, however, lost no time in deserting him, much to the indignation of Colonel Bouquet, who vowed he would cover them with shame by publishing their names in the local newspaper. It was on June 2 that the first demonstration of hostilities took place. Early in the morning a party of Indians, concealed by the woods and some outhouses, fired upon the fort. As the fire was innocuous, Blane husbanding his ammunition replied only by giving them 'three chears'; but as 'the poping' continued for many hours, at 5 P.M. he sent his sergeant with a small party to set the outhouses on fire—a measure which had the effect of driving the enemy away.

All now remained quiet until the 21st, when an attempt at drawing part of the garrison out into an ambush failed, and a serious attack was repelled. Blane was, however, not unnaturally anxious about his post, and more than once wrote to express a hope that Bouquet was marching to his relief. The line of his communication with his chief lay through Fort Bedford—50 miles distant—but letters seldom reached their destination, for the peril to his messengers can hardly be exaggerated.

'They were,' says Parkman, 'usually soldiers, sometimes backwoodsmen, and occasionally a friendly Indian, who . . . could pass when others would infallibly have perished. If white men, they were always mounted; and it may well be supposed that their horses did not lag by the way. The profound solitude; the silence, broken only by the moan of the wind, the caw of the crow, or the cry of some prowling tenant of the waste; the mystery of the verdant labyrinth, which the anxious wayfarer strained his eyes in vain to penetrate; the consciousness that in every thicket, behind every wall, might lurk a foe more fierce and subtle than the cougar or the lynx: and the long hours of darkness when, stretched on the cold ground, his excited fancy roamed in nightmare visions of a horror but too real and imminent,—such was the experience of many an unfortunate who never lived to tell it. If the messenger was an Indian, his greatest danger was from those who should have been his friends. Friendly Indians were told, whenever they approached a fort, to make themselves known by carrying green branches thrust into the muzzle of their guns; and an order was issued that the token should be respected. This gave them tolerable security as regarded soldiers, but not as regarded the enraged backwoodsmen, who would shoot without distinction at anything with a red skin.'

The enemy's attacks were happily less persistent than they had been elsewhere. Blane kept up good heart, and for the moment we may take leave of him still holding his post.

Fort Bedford

It remains only to tell the tale of Fort Bedford, the easternmost in the chain of posts which connected civilisation with the Great Lakes. At this spot, pleasantly situated on the slopes of the Alleghanies, Captain Ourry—apparently a Jersey man—beguiled with gardening the time that he could spare from his military duties in the garrison, which as we have already seen had consisted formerly of six, and more recently of five men. Bedford formed the halfway house between Carlisle and Fort Pitt: the former being ninety-five, the latter one hundred and five miles distant; and was the centre of numerous small settlements scattered amid the valleys. When the Indian outbreak took place, the families of refugees coming for shelter rapidly raised the number of the garrison to 125, out of which, in addition to the men of his own regiment, Ourry formed two companies of militia, a section of whom disguised as Indians proved, possibly for the first time in our service, the advantage of the mounted infantry arm; for they sallied out on horseback, took a party of savages by surprise, and drove them in headlong rout.

In consequence perhaps of this repulse the expected attack on the part of the enemy was delayed. The result was that the militia deserted Ourry, whose garrison was reduced by the middle of June to three corporals and nine privates of the Royal Americans, plus seven Indian prisoners—a somewhat unreliable addition. Yet this was all he had for the defence of a fort of five bastions! Then a sudden raid of the enemy brought back the refugees; but several had been murdered while mowing in the fields, and a letter from Fort Bedford dated June 30 states, 'Two men are brought in, alive, tomahawked and scalped more than half the head over. Our parade just now presents a scene of bloody and savage cruelty: three men . . . lying scalped thereon.'

The Red-skins continued their guerilla warfare, but declined to come to close quarters ; and Ourry, blockaded rather than besieged, held on resolutely awaiting the relief which was on its way.

CHAPTER VII

BOUQUET RELIEVES FORT BEDFORD—DEFEATS THE INDIANS AT BUSHY RUN—RELIEVES FORT PITT

It took long to convince Sir Jeffery Amherst of the perilous situation. To such an extent did he carry his contempt for the Indian tribes that he actually found fault with the officers in command of detached forts for taking the most ordinary precautions to strengthen their posts. Amherst's was a curious character: and in spite of the splendid results of his operations against Canada, it is difficult to believe him to have been a really great man. However, as far as our Regiment is concerned, we have every reason to be grateful for the care and interest which he took in all its concerns.[1]

In Amherst's private character we find a feature sometimes lacking in that of greater men. He could recognise the abilities of a clever subordinate and give him a free hand. In the present instance he gave his full confidence to Colonel Bouquet; and Bouquet, a man of far greater talent and breadth of view, who had realised the danger long before his chief had done so, repaid his confidence by unswerving loyalty, even when it is evident that their respective views did not coincide. His warnings had

[1] It is noticeable that when, in after years, he had become Commander-in-Chief of the British army, and presumably had plenty to occupy his time, his letters regarding the Regiment were always written in his own hand. Amherst in his later years became, like Lord Grenfell, Colonel of the Life Guards.

hitherto been unheeded by Sir Jeffery: but in the middle of June the latter, reinforcing Bouquet at Philadelphia with a company of the 42nd and one of the then 77th Highlanders, formulated his plan of campaign, under which Major Wilkins of the 60th was to proceed to the relief of Detroit by way of the Great Lakes; while Colonel Bouquet was to march in person by way of Fort Pitt to Presqu'île. Bouquet pointed out the need of abandoning the small outlying posts and concentrating their garrisons at Fort Pitt, but Amherst refused his consent. However on June 23 news of the massacre of a party of Englishmen roused the General to fury, and he took steps to reinforce Bouquet with the remainder of the 42nd and 77th Regiments; but the former comprised but 214, and the latter but 133, officers and men. Small as was this reinforcement it was the entire force available, for the residual fragment of the 17th Regiment had been already despatched to the Mohawk River, and a company of the 55th was under orders to proceed from Albany to Oswego. 'All the troops from hence that could be collected are sent you,' wrote Amherst to Bouquet on June 25, 'so that should the whole race of Indians take arms against us I can do no more.'[1] It was at this time that the General made the suggestion of inoculating the Indian tribes with small-pox. Our ideas have changed since those days, but one can hardly doubt that even at that time such a proposal would have been received by many with horror.

At the end of June Colonel Bouquet had crossed the Susquehanna and reached Carlisle, which he found crowded with refugees in such a condition of grief, misery, and terror that it required all his influence to stop the massacre

[1] Five weeks later the skeletons of five battalions from Cuba landed: but so great had been the ravages of fever that they numbered less than a thousand officers and men fit for duty.

of a few Indians, although known to be friendly, who had come to seek the protection of the soldiers. But the most curious part of the story is that neither the province of Pennsylvania nor the people on the exposed frontier would do anything for their self-preservation. Bouquet remarked that he hoped to be able to save the infatuated people from destruction, in spite of all their endeavours to defeat Amherst's measures. The fact was that the preponderating Quaker element in Pennsylvania which in past days had to its credit done something to ensure fair play for the Indians, was so much puffed up with a sense of self-satisfaction as to be unable to believe the stories of atrocities committed by its protégés, and it was not until the truth of these stories had been doubly and trebly proved, and it was impossible to ignore the danger further, that the most elementary measures were taken for the defence of the country. Even then the Quakers lost no opportunity of belauding the savages and belittling their own countrymen. Their confidence in their friends was, however, to some extent misplaced. 'I am a Quaker,' proudly remarked one who found himself face to face with an Indian. 'Are you indeed?' replied the latter, as he stretched him dead with a tomahawk!

With the frontiersmen the reason was different. They were convinced that a repetition of Braddock's defeat must take place, and their minds were set on flight could transport only be found. Bouquet's kindness of heart would not permit him to let this wretched multitude starve; and at Carlisle he was detained for eighteen days in the collection of transport and provisions. But for his tact, energy, and knowledge of the country the effort would probably have been vain.

At length he was able to move. But no wonder that

to the terror-stricken countrymen his task appeared to be an impossible one. His force included about 550 men—less than those actually killed with Braddock—and was composed largely of the two Highland regiments already mentioned, brave as lions, but some too sickly to march on foot, and none trained in the mysteries of Indian warfare. Indeed, in a letter to Amherst, the Colonel remarked that he could not trust a Highlander out of his sight without risk of losing him. Under these circumstances the more depended upon the contingent of Royal Americans—some 150 strong—those of the 1st Battalion being probably supplemented by some of the disbanded of the 3rd and 4th. Slowly the little army, impeded by its large convoy, crept along the Cumberland valley. Bouquet—too well versed in war to forget as more modern generals have done the absolute necessity of a commander marching with the advance-guard—walked ahead of his column, rifle in hand, at times actually taking the part of a scout.

The route selected was that on which General Forbes, at Bouquet's instance, had advanced to the capture of Fort Duquesne. Shippensburg, twenty-two miles west of Carlisle, was found to be full of a starving crowd. The steep slopes of the Alleghanies were painfully and wearily climbed; Forts Loudon and Littleton in turn reached. The latter was empty, for the Government of Pennsylvania, 'with incredible perversity'—to use the expression of the American historian Parkman—had refused to garrison it, and Bouquet was obliged to hand some of his precious ammunition and provisions to a few countrymen, in order to enable them to prevent its being burned. Juniata was also found to be empty: Ourry had wisely withdrawn the one or two men who guarded it, as a reinforcement to his own slender garrison. On the 25th Bedford, the frontier town of comparative civilisation, was reached. Ninety-five miles, since the start from Carlisle, had been marched in seven days;—under the circumstances a remarkable performance.

Warmly must the force have been welcomed by Louis Ourry and his refugees, no less than eighteen of whose neighbours had been scalped by the Indians; while, so tight had the cordon been drawn, that all communication with the outer world, even that of express riders, had been severed. The relief of Fort Ligonier, forty-nine miles distant, was, however, an even more pressing matter: the place was in great peril, although Ourry had reinforced Blane with twenty backwoodsmen who volunteered for the purpose; and the loss of the stores and ammunition lodged there might be fatal to Bouquet's plans. In this stress Bouquet sent forward thirty regular soldiers—no doubt Royal Americans—who by forced marches through the woods succeeded in reaching Ligonier without hindrance from the enemy until within sight of the fort.

By this time Colonel Bouquet had perfected his system of carrying on Indian warfare, and the following paragraphs give the substance of his methods in preparation, equipment, and tactics against the savage foe in his native forest. (*Vide* also Appendix IV.) It will be noticed that the idea of mounted infantry—a branch of the service in which our Regiment has had more recent experience—was mooted by our first Commanding Officer.

'It may be taken for granted in warfare with the Indians, 1st that their general maxim is to surround their enemy, 2nd that they fight in extended order and never in a compact body, 3rd that when attacked they never stand their ground, but immediately give way only to return to the charge when the attack ceases. These principles being admitted, it follows 1st that the troops destined to engage Indians must be lightly clothed, armed and accoutred; 2nd that having no resistance to encounter in the attack or defence they are not to be drawn up in close order—a formation which would only expose them to needless loss; 3rd, that all their evolutions must be performed with great rapidity, and the men enabled by constant practice to pursue the enemy closely when put to flight, and not to give him time to rally.'

Colonel Bouquet's suggestion was to form a battalion of 'hunters'—the exact equivalent of the word *Yäger*—500 strong, with two troops of light horse—in modern parlance, mounted infantry—and a company of artificers attached, composed of frontiersmen—from fifteen to twenty years of age—enlisted for fifteen years and specially trained for the service.

Clothing

The clothing of a soldier for the campaign should consist of a short jacket of brown cloth, a strong tanned shirt, short trowsers, leggings, moccassins, a sailor's hat, a knapsack for provisions and an oil surtout against the rain. This surtout should have two coats of oil, and with the second coat it will be useful to mix some dark greenish colour to make the coat less conspicuous in the woods.

Arms

Their arms, the best that can be procured, should be short fusils and some rifles with bayonets in the form of a dirk to serve for a knife; small hatchets, and leathern bottles for water.

Exercises

The soldiers before being armed must be taught to keep themselves clean and to dress in a soldier-like manner. The first thing they are to learn is to walk well, afterwards to run; and in order to excite emulation small prizes might from time to time be given to those who distinguish themselves. They must then run in ranks in extended order and wheel in that order; at first slowly, but by degrees with increasing speed. This evolution is difficult, but most important in order to fall unexpectedly in the flank of the enemy. The men are to disperse and rally at given signals; and particular colours should be given to each company as rallying points. The men must be trained to leap logs and ditches and to carry burdens proportionate to their strength.

When perfect in these exercises the young soldiers will receive their arms and follow the above-named evolutions on all kinds of ground. They will be taught to handle their arms with dexterity; and without losing time upon trifles to load and fire very quick, standing, kneeling or lying on the ground. They are to fire at a mark without a rest, and not allowed to be long in taking aim. Hunting and the award of small prizes will soon make them expert marksmen.

The men should learn to swim, pushing before them on a small raft their clothes, arms and ammunition; they must also learn to use snow-shoes; they must be taught to throw up entrenchments, make fascines and gabions; as well as to fell trees, saw planks, construct canoes, carts, ploughs, barrows, roofs, casks, batteaux and bridges; and to build ovens and log-houses. With practice the youngest among them will soon become tolerably good carpenters, masons, tailors, butchers, shoe-makers, etc.

Light Horse and Dogs

In order to complete the establishment of this Corps two troops of Light Horse, each composed of 50 men and Officers, should be attached thereto. These men are to perform the same exercises as the foot soldiers, but to be afterwards taught to ride and in especial to be very quick at mounting and dismounting with their arms in their hands. They must learn to gallop through the woods up and down hill, and to jump logs and ditches.

The horses must be strong and hardy and accustomed to feed in the woods; they must be thoroughly broken to fire, be practised in swimming rivers, etc. Saddles and accoutrements to be of the plainest description, strong and light. The mounted men to have the same equipment as those on foot, and armed with a short rifle and a battle axe with a long handle for use in case of the charge. Each of these mounted men should be provided with a bloodhound who would be of use in discovering the enemy's ambush, and following his tracks. They would seize the naked savages, or at all events give time for the horsemen to come up with them; they would also add to safety of the camp at night.

The men should take it in turns to go on hunting expeditions with their Officers and remain out of camp for some weeks at a time, taking with them a little flour, but otherwise relying on the game and fish caught.

Great care is to be taken to preserve purity of manners, order and decency among the men: this will be found much easier in the woods than in the neighbourhood of towns. It would be a good plan to give the men only a small portion of their pay in cash, the remainder will be accumulated for them until discharge; then they would receive the balance due to them and 200 acres of land.

Bouquet halted for three days at Bedford: then, protected by an advance-guard of riflemen, his little army, with infinite toil, scaled the main ridge of the Alleghanies, and on August 2 arrived without opposition at Fort Ligonier, and relieved Archibald Blane and his garrison.

At Bouquet's approach the Indians had vanished, and news of their whereabouts was not to be obtained. It was necessary to advance with redoubled caution: and, with a view to greater mobility, the Colonel determined to leave behind his ox wagons, reducing his transport which even then amounted to 350 pack-horses and a few oxen. On the 4th he resumed his march, and next morning started for the stream of Bushy Run, which he had fixed as the halting-place for the day; intending, after dark, to break up his bivouac and, in hope of avoiding an ambuscade, to traverse by night the perilous defile of Turtle Creek, where for some miles ran the track, flanked on one side by a wall of rock, on the other by a precipice with a river flowing below.

The heat of summer was intense; mosquitoes abounded; the water-springs were dry. Yet seventeen miles had been marched, and Bushy Run was but half a mile ahead, when, on a sudden, a volley from an unseen foe struck down the leading files of the vanguard.

The column at once closed up; a sergeant and two men still unwounded held their ground, and the advance-guard being reinforced had no great difficulty in beating off the assailants, and in following them up. But hardly had the pursuit ceased when the Indians returned to the attack;

and the firing was simultaneously taken up by other parties of savages concealed on the high ground overlooking the flank of the column. To gain these heights became imperative; and was effected by a charge of the whole line. But it was merely cutting through water. As soon as the savages had been dispersed on one side they appeared on another in greater numbers. Constantly reinforced the Indians succeeded at length in surrounding the column and in turning their attention to the convoy in rear. For the protection of the convoy the fighting line had no alternative but to fall back; and, every moment as it retired, the enemy's fire became hotter and more harassing. But, under these trying conditions, the steadiness of the British troops was conspicuous; and the convoy, cleverly posted, was regained. Attack after attack was repelled. After seven hours night put an end to the conflict.

But the prospect was the reverse of cheerful. Upwards of sixty men of his whole force, which in its entirety probably fell short of 500, had been struck. Bouquet's position upon high ground, from the tactical point of view fairly good, was waterless: and his men—particularly the wounded—and the horses were suffering agonies of thirst. After dark Bouquet retired a little to the hill where his baggage had been posted, and formed a ring round the convoy. For defensive purposes he was now better off, but there was still no water. Colonel Bouquet thoroughly realised the peril of his situation. In a despatch written that night to Sir Jeffery Amherst, after describing the action, he continues:

'We expect to begin again at daybreak. Whatever our fate may be, I thought it necessary to give your Excellency this early information, that you may at all events take such measures as you think proper with the provinces for their own safety and the effectual relief of Fort Pitt; as, in case of another engagement, I fear insurmountable difficulties in protecting and transporting our provisions, being already so much weakened by the losses of this day in men and horses; besides the additional necessity of carrying the wounded whose situation is truly deplorable.

'I cannot sufficiently . . . express my admiration for the cool and steady behaviour of the troops who did not fire a shot without orders, and drove the enemy from their posts with fixed bayonets. The conduct of the Officers is much above my praises.'

It needs but little reading between the lines to realise how critical Bouquet felt his position to be.

The night was passed in cold and darkness, for no fires could be permitted; but in the twilight of the following morning at 5 A.M. yells and cries from the surrounding bush at a distance of about 500 yards heralded another attack on the part of the Indians. Approaching under cover of the trees, the savages opened fire simultaneously on every side. Nearer and nearer they came—the perfection of skirmishers—and making a beautiful example of a mode of attack a century and a half in advance of its age. Then, when close at hand, they would make a rush, only to be repelled by the steady fire and levelled bayonets of the extended line of defenders. Again and again the same thing happened. Again and again the Royal Americans and Highlanders advanced to make a counter-attack, only to find their savage foes vanished into space. But each attack left the British force a little weaker, a little diminished in numbers; and, worse still, positively maddened by thirst. There could be no doubt that the action was going in favour of the Indians. For hours, the soldiers could, and unquestionably would, hold their ground, but their ultimate fate was no longer doubtful. The redoubled yells of the Indians, still concealed, but so close that Bouquet and his men could hear their derisive taunts, showed that the savages were well aware that their victory was at hand. It was impossible for the troops to quit either the wounded or the convoy; many of the horses had stampeded and their drivers had disappeared: and the immutable law that an immobile force must eventually be destroyed by a mobile one held good at that time not less than at the present day.

But, with the coolness of a man born to command, Colonel Bouquet was gauging the situation. There was still one chance. If the enemy could be allured into the open the tables would be turned, and the fire and bayonet of the regular troops would make short work of the Indian foe. But how was it to be done? Bouquet determined to solve the problem by the usually futile expedient of a feigned retreat. The moment was favourable, for the savages were becoming every moment more audacious. Bouquet ordered two companies of the Royal Americans to retire, and the gap thus made in the circle was filled by the extended files of the companies on either side. The orderly way in which these manœuvres were executed shows that the force was well in hand.

The Indians at once fell into the trap, dashed forward, and assailed the weakened line with great violence. Suddenly they found themselves attacked by the two companies above-named, which, having quitted the circle, and being concealed by a depression in the ground, now hurled themselves against the enemy's right flank. Taken as were the savages by surprise they met the unexpected assailants with great gallantry, but a bayonet-charge following up the volley quickly decided the matter, and the Indians fled in confusion. In their flight they crossed the line of fire of two companies (one of the 42nd, the other of the Royal Americans, the whole commanded by Captain Barrett of the latter) detailed as a support, and the rout was complete. The four British companies followed in hot pursuit, and the left of the Indians, although not attacked, seeing the fate of their colleagues followed their example, and fled in headlong confusion. One prisoner was taken. Our men were in no merciful mood, and he was immediately despatched.

The Indians left sixty dead upon the field, a large proportion of whom probably fell during the final charge. If, however, as seems pretty certain, this figure is exclusive of the number killed on the day previous, it speaks well for the skill of the British marksmen against an almost concealed and constantly moving foe. The opposing forces seem to have been numerically about equal. Bouquet's losses are stated at about five-and-twenty per cent. of his whole force, comprising 3 officers and 47 men killed, 5 officers and 54 men wounded, 5 men missing.[1] Among the officers wounded mortally was Lieutenant Dow of the Royal Americans, the D.A.Q.M.G. The loss of our Regiment individually was 7 killed and 5 wounded, the comparative immunity being no doubt due to the fact that our men had had greater experience than the Highlanders in taking cover.

Bouquet lost no time in following up his victory. The same afternoon, as soon as litters could be improvised for the wounded, he pushed on to Bushy Run. Hardly had he formed his camp when, to the astonishment of all, they were again fired upon by the Indians. The bold assailants were, however, at once driven off by the Royal Americans, and no further incident occurred that day. But the force was unfortunately weakened by the fact that a quantity of horses had galloped off during the action, and it had in consequence become necessary to destroy such part of the provisions as could no longer be carried.

Fort Pitt was now about twenty-five miles distant.

[1] Parkman raises the number of casualties to 123.

Colonel Bouquet resumed his march on the 7th, but was so much harassed by his pertinacious foe that it was not until the 10th that he was able to reach his destination and relieve Ecuyer and the garrison. The losses both of men and material, which he had incurred *en route*, made it impossible for the force to cross the Ohio for its ulterior object of relieving Detroit. Bouquet's march of 324 miles from Lancaster was therefore at an end, and he perforce contented himself with provisioning Fort Pitt and the other stations on the line of communication, and making preparations for winter quarters.

His operations had, it is true, not put an end to the rebellion, but they had filled the Indians with consternation; and the women, by the signs of grief and mourning prevalent among barbarians, showed how thoroughly they appreciated the importance of their husbands' defeat. Sir Jeffery Amherst was most complimentary to the victor. The news was received in the provinces with transports of joy and admiration; those acquainted with the difficulties of Red Indian warfare being the loudest in their praise. The Assembly of Pennsylvania, to its credit, passed a vote of thanks to Colonel Bouquet. And in truth the battle of Bushy Run forms a landmark in the history of our Empire. It was not merely that so sternly contested an action had never previously been fought between Briton and Red-skin, but that Braddock's dying prophecy had been carried out in its entirety; we had 'learned to do it better next time.' Yet it may be doubted whether Bouquet was not the one and only man in America who could have won that victory; and it will be noticed that the force which conquered at the crisis was one not of provincial levies but of the Royal Army.

Since August 5 and 6, 1763, the 42nd and the 60th

have stood side by side on many a hard-fought field. Splendid as is the record of the Highlanders, it is possible that there may be no action in which their steadfast gallantry has been put to a higher test, or on which they may look back with greater satisfaction than the fight at Bushy Run, led by their heroic Major, Allen Campbell, and under the command of a Swiss Colonel whom his adopted country has long ago forgotten.

(From an engraving after the portrait by Benjamin West, P.R.A.)

CHAPTER VIII

SIEGE OF DETROIT CONTINUED—DEATH OF CAPTAIN CAMPBELL—ESCAPE OF ENSIGN PAULI—EXPLOIT OF SERGEANT SMITH—PONTIAC ABANDONS THE SIEGE—FAILURE OF MAJOR WILKINS—SIR J. AMHERST IS SUCCEEDED BY GENERAL GAGE—THE DEVIL'S HOLE—COLONEL BOUQUET AND COLONEL BRADSTREET DIRECTED TO ENFORCE THE SUBMISSION OF THE INDIANS—BRADSTREET'S COMPLETE FAILURE—BOUQUET'S ADVANCE TO THE MUSKINGUM—HIS RESCUE OF THE PRISONERS—GRATITUDE OF THE WHOLE COUNTRY—BOUQUET'S PROMOTION, AND DEATH

WE must now return to the garrison of Detroit which we left, as described in Macdonald's graphic letter, beleagured by a force of some 600 Indians. Major Gladwyn believed that the revolt would subside as quickly as it had sprung up; and being, moreover, in want of provisions, determined to open negotiations with Pontiac. As a result Pontiac requested that Captain Donald Campbell of the Royal Americans, second-in-command to Gladwyn, and a *persona grata* to the Indians, should visit their camp. Gladwyn was far from satisfied that treachery was not intended; but Campbell, who, as far as we can judge from the very meagre account of him, appears to have been a man of fine character and liked by the Indians, begged to be allowed to hazard his life for his friends. Gladwyn reluctantly consented, and Campbell, accompanied by Lieutenant Macdougal of his regiment, walked off to the Indian camp. The reception of these officers was anything but reassuring; and in truth, but for the personal intervention of Pontiac they might have been murdered on the spot; even as it was, a very short space of time sufficed to show them that they were prisoners in the hands of the savages.

Gladwyn's difficulties were by no means confined to the conflict with the Indians. He was surrounded by Canadians, of whom a large proportion was actively hostile and prepared to go any lengths in concert with the Indian tribes to compass the destruction of the British forces. Some, however, adopted a neutral attitude; a few were friendly; and even among the Indians themselves, many either joined Pontiac under compulsion or refused to join him at all—a decision due for the most part to the efforts of the Jesuit priests who utterly declined to join their countrymen in anything which could tend to the destruction of civilisation by barbarism.

As early as May 11, Gladwyn had consulted his colleagues. Some [1] we are told, said that the case was desperate, and the sole resource to embark on the lake and sail to Niagara. Relief could not be expected within three weeks; yet three weeks was the longest period for which they could hope to eke out their scanty store of provisions. The straw-thatched houses, too, lent themselves to conflagration from burning arrows. To his credit, the Commandant refused to listen to any such counsel: he was resolved to hold his post or to die therein. By clearing the ground in front of the fort, a measure needing the services of volunteers, Gladwyn to a great extent deprived the enemy of cover, while the two vessels in the river enfiladed the ground immediately to the north

[1] Macdonald was evidently not one of them. *Vide* his letter, p. 137.

and south of the fort. Even so the savages, most skilful of skirmishers, would worm their way through the grass almost to the palisade, and shoot arrows tipped with burning tow upon the houses. The position of the church was one of special danger, for it was greatly exposed; and if burned to the ground would have left a fatal gap in the defences. But here again the Jesuit priests afforded material aid by threatening Pontiac with the vengeance of the Great Spirit should he commit such sacrilege. Thus time passed on; and the Indians, finding their efforts unavailing, gave up the idea of close attack and contented themselves with harassing the garrison from a distance. But the question of provisions must have decided the fate of Detroit had it not been for the aid of M. Baby, one of the few friendly Canadians living on the opposite side of the river, who under cover of night by means of boats furnished the garrison with a fairly plentiful supply of meat and other necessaries.

The days went by, but the vigilance of the garrison never relaxed. It was impossible for a soldier even to show his head above the palisade without becoming a target for the Indian sharpshooters. But no succour appeared; and Major Gladwyn despatched one of his ships to Niagara with orders to hasten the sorely needed aid. Next day the vessel became becalmed; a storm of canoes darted out to attack her, and in the prow of the first of these, the dastardly savages had placed Captain Campbell. But to the honour of himself and his regiment, the gallant officer called out to the crew to fire upon the Indians regardless of what might happen to him. Fortunately at the moment a breeze sprang up, and the schooner escaped its disappointed foes.

'The Fort, or rather town of Detroit,' says Parkman, 'had by this time lost its wonted vivacity and life. Its narrow streets were gloomy and silent. Here and there strolled a Canadian in red cap and gawdy sash ; the weary sentinel walked to and fro before the Quarters of the Commandant ; an Officer, perhaps, passed along with rapid step and anxious face ; or an Indian girl, the mate of some soldier or trader, moved silently by in her finery of beads and vermilion. Such an aspect as this the town must have presented on the morning of the 30th of May, when, at about 9 o'clock, the voice of the sentinel sounded from the South-east bastion ; a loud exclamation in the direction of the river roused Detroit from its lethargy. Instantly the place was astir. Soldiers, traders and *habitants*, hurrying through the watergate, thronged the canoe wharf and the narrow streets without. Half wild *coureurs de bois*, the tall sinewy provincials, and the stately British soldiers, stood crowded together, their uniforms soiled and worn, and their faces haggard with unremitted watching. Yet all alike wore an animated and joyous look. The long expected convoy was in sight. On the far side of the river, at some distance below the Fort, a line of boats was rounding the woody projection called Montreal Point, their oars flashing in the sun, and the red flag of England flying from the stern of the foremost. The toil and dangers were drawing to an end. With one accord, they broke into three hearty cheers, again and again repeated, while a cannon, glancing from the bastion, sent the loud voice of defiance to the enemy, and welcome to approaching friends. But suddenly every cheek grew pale with horror. Dark naked figures were seen rising with wild gestures in the boats ; while, in place of the answering salute, the distant yell of the war whoop fell faintly on their ears. The convoy was in the hands of the enemy. Officers and men stood gazing in mournful silence, when an incident occurred which caused them to forget the general calamity in the absorbing interest of the moment.

'. . . In each of the boats, of which there were eighteen, two or more of the soldiers, deprived of their weapons, were compelled to act as rowers, guarded by several armed savages, while many other Indians, for the sake of further security, followed the boats along the shore. In the foremost, as it happened, there were four soldiers and only three Indians. The larger of the two vessels still lay anchored in the stream about a bow-shot from the Fort, while her companion, as we have seen, had gone down to Niagara to hasten up this very reinforcement. As the boat came opposite this vessel, the soldier who acted as steersman conceived a daring plan of escape. The principal Indian sat immediately in front of another of the soldiers. The steersman called in English to his comrade to seize the savage and throw him overboard. The man answered that he was not strong enough ; on which the steersman directed him to change places with him as if fatigued with rowing, a movement which would excite no suspicion on the part of the guard. As the bold soldier stepped forward as if to take his companion's oar, he suddenly seized the Indian by the hair, and gripping with the other hand the girdle at his waist, lifted him by main force and flung into him the river. The boat rocked till the water surged over her gunwale. The Indian held fast to his enemy's clothes, and drawing himself upward as he trailed alongside, stabbed him again and again with his knife, and then dragged him overboard. Both went down the swift current, rising and sinking, and, as some relate, perished grappled in each others arms. The two remaining Indians leapt out of the boat. The prisoners turned and pulled for the distant vessel, shouting aloud for aid. The Indians on shore opened a heavy fire upon them, and many canoes paddled swiftly in pursuit. The men strained with desperate strength. A fate of inexpressible horror was the alternative. The bullets hissed thickly around their heads ; one of them was soon wounded, and the light birch bark canoes gained upon them with fearful rapidity. Escape seemed hopeless, when the report of a cannon burst from the side of the vessel. The ball flew close past the boat, beating the water in a line of foam, narrowly missing the foremost canoe. At this the pursuers drew back in dismay ; and the Indians on shore being farther saluted by a second shot, ceased firing and scattered among the bushes. The prisoners soon reached the vessel, where they were greeted as men snatched from the jaws of fate, "a living monument," writes an Officer of the garrison, "that fortune favours the brave."'

The story they had to tell was one of misfortune. As early as May 13 they had left Fort Niagara under the command of Lieutenant Cuyler of the 55th Regiment. His party consisted of ninety-six men. During the 28th he had reached the mouth of the Detroit River, and disembarked to encamp. On a sudden they were surprised by the Indians: the party rushed down to the beach, attempting to escape in boats, but all were captured with the exception of two boat-loads. It was these prisoners who were now being conveyed by water to Pontiac's camp, where they were received with a wild orgy of drunkenness, during the course of which the prisoners, sixty-six in number, were murdered with every circumstance of horror. Rumours of what had happened reached the fort the same night from the reports of Canadians.

> 'On the following day,' says Parkman, 'and for several succeeding days, they beheld frightful confirmation of the rumours they had heard. Naked corpses, gashed with knives and scorched with fire, floated down the pure waters of the Detroit, whose fish came up to nibble at the clotted blood that clung to their ghastly faces.'

And now almost day by day came news of the loss of British posts on the Great Lakes; Sandusky, Presqu'île, Michilimackinac and the rest. But on June 23, Gladwyn's schooner, returning from Niagara, brought a welcome reinforcement of men, provisions, and ammunition; and not only this, but having on the way inveigled the Indians into attacking them with their canoes, had given the savages a much-needed lesson with grape and musket-shot. The schooner also brought news that peace had been concluded between France and England; and that the former had ceded Canada with the region of the Lakes to Great Britain.[1] The position of the Canadians was therefore changed. They had to realise that they were now subjects of the British Crown. Furious at the tidings, several at once openly joined Pontiac in his camp; where they were received with welcome not unmingled with contempt, and regaled with an appetising feast of dogs' flesh. After the banquet a couple of Indians intoxicated with rum ran out towards the fort, boasting of their prowess. Two rifle-shots rang out, and the warriors fell dead upon the ground.

On the following night in proof of *bona fides* the Cana-

[1] This must have been already known in all but the most outlying places.

dians, accompanied by some of their new friends, advanced against the fort and established themselves in an entrenchment close by. Their advantage was not destined to be of long duration. At daybreak Lieutenant Hay of the Royal Americans with a file of men, sallied out to dislodge them. Two Indians were killed, but the Canadians fled with such rapidity that they all escaped unhurt.

It unfortunately happened that one of the soldiers with Hay had lived for many years among the Delawares as a prisoner, and had acquired some of their customs and habits; it was therefore the most natural thing in the world to him to take the scalp from one of the dead Indians. Still more unfortunately it happened that the savage in question was a nephew of the Chief of the Ojibwas; and hearing of his relative's death, the enraged chief seized Captain Campbell, and had him instantly shot to death with arrows. He then devoured his heart, a practice considered to be an exceptional honour paid only to a foe of acknowledged bravery. The feelings with which Campbell was regarded are shown in a passage of Macdonald's letter quoted above.

> 'I never have nor ever shall have a friend or acquaintance that I value more than him. My present comfort is that if charity, benevolence, innocence, and integrity are a sufficient dispensation for all mankind they entitle him to happiness in the world to come.'

His dying agonies were soothed with the reflection that he had sacrificed himself for his friends,

> *He died with a smile upon his face, and glory in his soul.*

Happily, Campbell's subaltern Macdougal had already escaped.

The incidents of the day were, however, by no means over. During the afternoon an Indian, hotly pursued by his comrades, was seen making his way towards the Fort. The gates were opened to receive him, and he fell panting upon the ground; when, to the surprise of all, it was found that the *soi-disant* Indian was in reality Ensign Pauli of the Royal American Regiment. On regaining breath Pauli related his story. After the destruction of Sandusky, being reserved by his captors for torture, he had been brought into Pontiac's camp; but as fate would have it, an old squaw who had recently lost her husband, fell in love at first sight with the officer, and immediately proposed marriage. Pauli with pardonable alacrity consented; and, having been ceremonially ducked in the lake to eliminate all stains of white blood, for the next few weeks resided among the savages with the honours usually accorded to a chief. It may, however, be presumed that his domestic felicity was not entirely unchequered; and whatever may be the opinion of the strict moralist, it is not altogether matter for wonder that the Ensign had taken the earliest opportunity to escape from the embraces of his bride. Headed by Mrs. Pauli who, tomahawk in hand, was bent on vindicating the rights of her sex with a vigour hardly exceeded by the most militant suffragette of the present day, the tribe followed in hot pursuit, and Pauli barely escaped with his life.

Two of the Indian tribes, the Wyandots and Pottawattamies, had by this time begun to tire of the siege, and shortly afterwards gave up to Gladwyn the whole of the white prisoners in exchange for a treaty of peace. At the end of July a reinforcement of 280 men from the 55th and 80th Regiments arrived from Niagara under Captain Dalzell, A.D.C. to Sir Jeffery Amherst. At the urgent request of this officer, the Commandant rather reluctantly gave him permission to attack the camp of the Indians on the following night. What had happened so often before now occurred again. Dalzell fell into an ambuscade. The party returned to the fort with a loss of fifty-nine killed and wounded—one man R.A.R. killed: six wounded—among the killed being Dalzell himself; and the repulse of Bloody Bridge is not forgotten at the present day. August passed away marked by no incident other than an occasional skirmish; but on September 4, the smaller of Gladwyn's schooners, which had been sent to Niagara with the despatches, cast anchor on her return voyage about nine miles below the fort. The night was pitch dark; and under cover of it, despite the ceaseless vigilance of the crew, 350 Indians glided down the stream in their canoes. So close were they when detected that there was only time to fire one cannon-shot. The next instant, with their knives between their teeth, the savages were clambering up the vessel's sides unchecked by a volley of musketry, and placed only at a momentary disadvantage by a furious charge of the ship's crew armed with spears and hatchets. Quickly recovering themselves in spite of the loss of a score of their comrades, the Red-skins once more leapt over the bulwarks, killed the master and disabled several of the crew. At this desperate moment when all seemed lost, the situation was saved by Sergeant Smith of the Royal Americans, who leaped down the hatchway, lighted match in hand, loudly proclaiming his intention of blowing up the vessel. Some of the savages grasped his meaning; and the whole jumping overboard, swam away in every direction. Next morning the schooner entered the harbour of Detroit, Smith and his companions covered with blood from the wounds inflicted by the scalping knives.[1]

[1] It is only fair to say that Parkman gives the credit of this exploit to Jacobs, the mate; but Patrick Murray is emphatic in saying that Smith was the hero of the episode, adding moreover that it was Smith who brought the vessel into harbour.

The schooner had brought a supply of provisions, welcome indeed, but inadequate for the requirements of the garrison, which continued to be placed on the smallest possible allowance. Thus the month of September passed away. The Indians, disappointed of their prey, were rapidly tiring of the siege which they had prosecuted with a vigour unprecedented in their annals. News arrived that Major Wilkins was on his way from Niagara, and they decided that the best plan of action would be to profess penitence, beg for peace, and renew the war unexpectedly during the following spring. With this object, the representatives of all the tribes, excepting the Ottawas, sought an interview with Major Gladwyn. The latter was not for a moment deceived, but cleverly turned the situation to good account. Peace he said he was not empowered to conclude; but granted a truce; and under cover of it scoured the surrounding country, and collected such a supply of provisions as would enable him to subsist through the winter. Pontiac and his Ottawas still kept up a guerilla warfare.

Peace between France and England had, as already stated, been declared early in the year. Amherst, on hearing of it, had lost no time in sending the news to M. Neyon de la Vallière, the French Commandant of Fort Chartres, the principal post in the Illinois district, with a request that it might be communicated to the Indian tribes friendly to his country. Much to his disgust, Neyon felt compelled to comply; and on October 31 Pontiac received from him a warning that no assistance could be hoped for from the French, and that the Indians must make the best terms they could with their British foe. This letter was the death blow to Pontiac's hopes. For the time being, at all events, his cause was lost; but with profound dissimulation the chief concealed his anger, and straightway wrote a letter in French to Gladwyn, saying that he had buried the hatchet, and requesting him to forget the past. Pontiac then withdrew to the Maumee with the intention of stirring up the Indians in that neighbourhood, and renewing hostilities after the winter.

The siege of Detroit was at an end. For five and a half months Gladwyn and his garrison had stood with unremitting vigilance the strain of alternate hope and fear, hunger, repeated disappointment, and incessant toil, chequered by disaster which even now was not at an end, for hardly had Pontiac retired when a messenger arrived with a letter saying that Major Wilkins of the Royal Americans had started from Niagara with a detachment; that many of his boats had been wrecked, and seventy of his men had perished in a storm; that the whole of his stores and ammunition were lost, and the expedition forced to return to its starting-point. With this dreary news the men of Detroit sat down to face the rigours of the coming winter.

During the autumn of the year, Sir Jeffery Amherst resigned his command, handing it over to General Gage, but retaining the Colonelcy-in-Chief of the Royal Americans. The last year had not increased Amherst's reputation; but his loss was soon felt, for his successor was a man of less ability.

It will be remembered that Sir Jeffery's plan for the relief of Detroit had included the co-operation of Bouquet's force with a detachment from the garrison of Niagara, a fort which the Indians had found too strong to be attacked. With this intent a quantity of military stores were, early in September, conveyed by boat up the Niagara River and landed at the site of the modern Lewiston. Thence, on the 13th, they were carried over a *portage* road several miles long by wagons and pack-horses to Fort Schlosser, a short distance above the celebrated falls. Next day the wagons and horses, guarded by an escort of twenty-four soldiers, set out on its return journey.

'The road,' observes Parkman, 'traversed a region whose sublime features have gained for it a world-wide renown. The River Niagara a short distance below the cataract assumes an aspect scarcely less remarkable than that stupendous scene itself. Its channel is formed by a vast ravine; whose sides, now bare and weather-stained, now shaggy with forest-trees, rise in cliffs of appalling height and steepness. Along the chasm pour all the waters of the lakes, heaving their furious surges with the power of an ocean and the rage of a mountain torrent. About three miles below the cataract, the precipices which form the eastern wall of the ravine are broken by an abyss of awful depth and blackness, bearing at the present day the name of the Devil's Hole. In its shallowest part, the precipice sinks sheer down to the depth of eighty feet, where it meets a chaotic mass of rocks, descending with an abrupt declivity to unseen depths below. Within the cold and damp recesses of the gulf, a host of forest-trees have rooted themselves; and, standing on the perilous brink, one may look down upon the mingled foliage of ash, poplar, and maple, while above them all, the spruce and fir shoot their sharp and rigid spires upward into sunlight. The roar of the convulsed river swells heavily upon the ear; and, far below, its headlong waters, careering in foam, may be discerned through the openings of the matted foliage.'

Even with the wagons unladen, the progress of the party was slow; but at length it reached the point 'where the road passed along the brink of the Devil's Hole. The gulf yawned on their left, while on their right the road was skirted by low densely wooded hills. Suddenly they were greeted by the blaze and clatter of a hundred rifles. Then followed the startled cries of men, and the bounding of maddened horses. At the next instant, a host of Indians broke screeching from the woods, and rifle and tomahawk finished the bloody work. All was over in a moment. Horses leaped the precipice; men were driven shrieking into the abyss; teams and wagons went over, crashing to atoms among the rocks below.' The conductor of the convoy, being well mounted, charged back through the Indians, and regained Fort Schlosser. Of the remainder, all but two perished; one being a teamster who lay concealed in the bushes; the other, a drummer boy, whose drumstrap, as he fell over the precipice, caught in the branches of a tree, from which he eventually disentangled himself.

The sound of firearms had been heard by a small detachment left to guard the lower landing place. The men at once advanced to the relief of their comrades. More haste, less speed. Anticipating the movement, the Indians had formed an ambuscade; and as the eager party advanced, struck half the detachment with a single volley and cut down almost the whole of the remainder. Ten only—eight of them wounded—who survived, fled to Niagara; whence Major Wilkins, the Commandant, lost no time in marching with a strong force of his regiment to the scene of the massacre. Wilkins and his men found that the Indians had disappeared, but the scalped and mangled corpses of their comrades strewed the ground. No less than eighty-one had been killed. The Royal Americans were not mixed up in this disastrous affair, except to the extent of losing three men wounded.

Several weeks elapsed before Major Wilkins, at the head of a force of six hundred regular soldiers, collected from far and wide, was able to make another effort for the relief of Detroit. But fortune was still untoward. As his boats were slowly making headway against the swift current above Niagara Falls, they were set upon and taken at a disadvantage by a small body of Indians, and driven back with considerable loss to Fort Schlosser. Once more he made the attempt, and reached Lake Erie; but, when at no great distance from Detroit, was overtaken at night by a violent storm. Of his boats many foundered: their crews perished, their stores sank, and the remains of the flotilla, as already noticed, had no resource but to return to Niagara.

The autumn and winter of 1763 passed away drearily enough for the outlying detachments. Writing a few days after the action of Bushy Run, Lieutenant Blane, after congratulating Colonel Bouquet on his victory, begs, as he expresses it, that he may not be left forlorn but at once be given an adequate garrison. In another letter he points out that there is neither surgeon nor medicine for the sick, and that his 'poor starved militia' were destitute even of shirts and shoes. Captain Ecuyer, writing to Bouquet on November 13, humorous even in his annoyance, adds his tale of complaint: desertion was rife, and insubordination rampant. Bouquet diagnosed the case quickly enough: he attributed the unprecedented desertion to the orders of Amherst, who had kept the battalion for seven years in the wild forests far from civilisation, and had forbidden the discharge of time-expired men. It is perhaps then hardly matter for surprise, that the men, deprived of their legal rights, took the matter into their own hands and yielded to the temptation of desertion, confident of being screened and sheltered by the sympathy of the settlers. So deeply was Bouquet affected that he begged to be relieved of his command—a request which General Gage with very proper judgment refused. The General had, in fact, a task for him of the utmost importance, namely that of advancing into and beyond the valley of the Ohio, and enforcing obedience on the recalcitrant Indian tribes. Concurrently with this advance, Gage directed a second force under command of another Royal American officer, the well-known Colonel Bradstreet, to sail up the lake and reduce to submission the Algonquins at Detroit and the regions beyond.

Bouquet's task bristled with difficulties. He had to penetrate interminable forests and unexplored country, destitute of roads, and uninhabited by civilised man. Every single article, whether of ammunition, stores, or provisions, had to accompany the column. In the event of failure he had no line of retreat. The troops detailed for the service comprised two very weak battalions, namely the 42nd Highlanders and the remnant of the 1st Battalion of the Royal Americans. Provincial troops were expected from Virginia and Pennsylvania, and two hundred friendly Indians were engaged. But the Indians never came: the Government of Virginia said it could not raise men; and the Pennsylvanian contingent of one thousand did not arrive until the beginning of August 1764. On the 5th of that month Bouquet assumed command of the whole at Carlisle, a town one hundred and eighty miles west of Philadelphia. The next four days were spent in necessary preparations, in organising the force, in laying down strict regulations for its discipline, and in cutting down the baggage to a minimum. On the 13th Bouquet reached Fort Loudon, the force being already reduced by the desertion of three hundred of the Pennsylvanians. The Provincial Government gave orders, however, that the deserters should be replaced.

At Loudon Bouquet was astonished to receive a communication from Colonel Bradstreet to the effect that he might return home with his troops, as a treaty had been concluded with the Delawares and the Shawanoes, two of the tribes whom Bouquet—Bradstreet's own Colonel, be it remembered—had been especially despatched to reduce to submission. It appeared that Bradstreet, having concentrated his force—for the most part provincials—at Albany, had set out thence about the end of June, and moving by the usual military road had ascended the valley of the Mohawk, crossed the Oneida Lake, and descended the Onondaga. His boats reaching Oswego, and setting forth on Lake Ontario—names which in the Indian tongue mean respectively Rapid River and Pretty Lake—soon encountered a storm which threw the flotilla into confusion. Niagara was at length reached, and a camp formed for the twelve hundred men composing the expedition. It so happened that an enormous concourse of Indian tribes—Menominies, Ottawas, Ojibwas, Wyandots, Iroquois—to the number of over two thousand men had, at the invitation of Sir William Johnson, already arrived at Niagara. Although the meeting was one of friendship, the Indians frequently took the opportunity of shooting any straggling soldiers; and every precaution that vigilance could ensure had to be taken by the Niagara garrison. The efforts of Sir William Johnson were on the whole fairly successful; the Indians departed, and Bradstreet pushed on to Fort Schlosser. Continuing on July 8 his course up Lake Erie, he disembarked at Presqu'île on the 12th and formed his camp. News arrived that strange Indians were at hand; and a party of savages, claiming to be Chiefs of Delawares and Shawanoes empowered to treat for peace, appeared upon the scene.

In spite of all warnings that the ambassadors were lacking in the necessary credentials; in spite of the dissent of his officers and the wrath of his Indian allies, Bradstreet straightway concluded with them a preliminary treaty, promising to refrain from attacking the two tribes on condition that the deputies should meet him again at Sandusky in twenty-five days' time, give up their prisoners, and conclude a definite peace. He then wrote to Colonel Bouquet in the terms already mentioned.

Bouquet naturally enough was much incensed, and vented his feelings in a letter to the Commander-in-Chief:

'The terms he [Bradstreet] gives them are such as to fill me with astonishment. . . . Had Colonel Bradstreet been as well informed as I am of the horrid perfidies of the Delawares and the Shawanoes, whose parties as late as the 22nd inst. killed six men . . . he never could have compromised the honour of the nation by such disgraceful conditions, and that at a time when two armies, after long struggles, are in full motion to penetrate into the heart of the enemy's country. Permit me likewise, humbly to represent to your Excellency that I have not deserved the affront laid upon me by this treaty of peace (concluded by a younger officer, in the department where you have done me the honour to appoint me to command), without referring the deputies of the savages to me at Fort Pitt, but telling them he shall send and prevent my proceeding against them. I can therefore take no notice of his peace, but proceed forthwith to the Ohio where I shall wait till I receive your orders.'

General Gage was quite equal to the occasion. He sent to Bradstreet a letter of severe reprimand, and to Sir William Johnson one expressing his astonishment and annoyance. In order to emphasise his condemnation of Bradstreet's breach of etiquette the General sent both these missives to that officer through Colonel Bouquet. Meanwhile, Bradstreet had resumed his progress, and reached Sandusky, where he again disobeyed his orders; and instead of attacking the Wyandots, Miamis, and Ottawas, allowed himself to be taken in a second time by acceding to their proposals to follow him to Detroit, and there conclude a treaty. On August 26, his force landed at Detroit amid the salute of cannon and the cheers of the garrison. The siege, indeed, had long come to an end; nevertheless the Indians were perpetually on the look out, and no soldier could enter the woods but at the peril of his life. Winter had passed in comparative quiet, but spring brought a renewal of hostilities, not indeed, as strenuous as those of the year before, but enough to harass the garrison, and to impose on it nights and days of ceaseless vigilance.

'Cut off,' says Parkman, 'for months together from all communication with their race; penned up in irksome imprisonment, ill supplied with provisions, and with clothing worn threadbare, they hailed with delight the prospect of a return to the world from which they had been banished so long. The army had no sooner landed than the garrison was relieved, and fresh troops substituted in their place.'

Of the Canadians who had given aid to the enemy, the greater part had fled; the remainder were tried, but a few only were found guilty and punished. Bradstreet's next course was to summon the Indians in council to meet him on September 7, when he once more allowed them to throw dust in his eyes, contenting himself with the most superficial professions of submission, while at the same time behaving with so little tact as to give great offence to the assembled tribes. Having concluded a treaty, he sent a detachment to repossess itself of Michilimackinac, Green Bay, and Sault Ste. Marie.

Then leaving Detroit, Bradstreet returned to Sandusky, where he not only received General Gage's letter of reprimand, but realised for the first time how thoroughly he had been duped. In utter disregard of their promises, the Delawares failed to bring in a single prisoner. Gage had ordered him to attack without delay the Indians on the plains of the Scioto. He utterly refused to obey, remained idling at Sandusky for several days, then broke up his command with such haste as to leave behind two of his men whom he had sent out to fish in the lake. On his return voyage, through lack of the simplest precautions he lost half of his boats and six guns. A part of his force was, consequently, obliged to return overland to Niagara, and many men, after suffering the extremity of hardship, died by the way. The main body reached the fort on November 4, and subsequently went into winter quarters at Oswego.

It is now time to return to Colonel Bouquet whom we left in August at Fort Loudon. On September 17 he reached Fort Pitt, and at the end of the month was joined by a serviceable contingent of Virginians whom the exertions of an intimate friend had attracted to his standard. Thus reinforced the Colonel left Fort Pitt on October 2 with a body of 1500 men, including drivers and other necessary attendants. Discipline had been restored and two soldiers shot for desertion. Of the women, one per corps and two hospital nurses were allowed to remain with the army. The rest had been sent to the domain of civilisation.

Bouquet's dispositions for the march were made as follows. The Virginian volunteers furnished an advance-guard, divided into three parties following three parallel paths; these were followed by pioneers and two companies of light infantry also formed into three detachments, under the direction of the Chief Engineer, with a view to securing the three paths. The column in the rear was so disposed as to be able readily to form a square in the presence of the enemy. The front face was formed by part of the 42nd; the left face, on the left-hand path, was made up of a battalion of Pennsylvanians; and the right face, marching by the right-hand path, consisted of the Royal Americans and the rest of the 42nd in single file. The convoy marched in the centre, while another battalion of Pennsylvanians formed the rear face. The strictest silence was enjoined, and the men marched in extended order at intervals of two yards from each other. In the event of attack they were directed to face outwards and form the square.

'Encumbered with their camp equipment, with droves of cattle and sheep for subsistence, and a long train of pack-horses laden with provisions, their progress was tedious and difficult, and seven or eight miles were the ordinary measure of a day's march. The woodsmen of Virginia, veteran hunters and Indian-fighters, were thrown far out in front and on either flank, scouring the forest to detect any sign of a lurking ambuscade. The pioneers toiled in the van; while the army dragged its weary length behind them through the forest, like a serpent creeping through tall grass. The surrounding country, wherever a casual opening in the matted foliage gave a glimpse of its features, disclosed scenery of wild, primeval beauty. Sometimes the army defiled along the margin of the Ohio, by its broad eddying current and the bright landscape of its shores. Sometimes they descended into the thickest gloom of the woods, damp, still, and cool as the recesses of a cavern, where the black soil oozed beneath the tread, where the rough columns of the forest seemed to exude a clammy sweat, and the slimy mosses were trickling with moisture; while the carcasses of prostrate trees, green with the decay of a century, sank into pulp at the lightest pressure of the foot. More frequently, the forest was of a fresher growth; and the restless leaves of the young maples and basswood shook down spots of sunlight on the marching columns. Sometimes they waded the clear current of a stream, with its vistas of arching foliage and sparkling water. There were intervals, but these were rare, when, escaping for a moment from the labyrinth of woods, they emerged into the light of an open meadow, rich with herbage, and girdled by a zone of forest; gladdened by the notes of birds, and enlivened, it may be, by grazing herds of deer. These spots, welcome to the forest traveller as an oasis to a wanderer in the desert, form the precursors of the prairies, which, growing wider and more frequent as one advances westward, expand at last into the boundless plains beyond the Mississippi.'

Logstown, between seventeen and eighteen miles from Fort Pitt, was passed on October 5, and the Muskingum near Tuscaroway, about ninety-three miles from Fort Pitt, was reached on the 13th. The river is here about seventy yards broad. Next day (Sunday) the army halted, and a report reached Bouquet that the Delawares and Shawanoes were coming to ask for peace. On the day following, this rumour was confirmed by the arrival of a deputation of Indians. The Colonel, having ordered a small stockaded fort to be built, consented to meet the savages. During the 17th, on the bank of the Muskingum under a roughly constructed arbour formed by boughs of trees, the council took place. The Indian deputation duly appeared: it included Kiashuta and Custologa, the representative chiefs of the Senecas and Delawares; Keissinaulchtha, the pride of the Shawanoes; Beaver, another Delaware—each of whom brought from six to twenty warriors. The scene was remarkable: the troops were drawn up in a meadow opposite, where glittering bayonets and the ordered ranks extending in long lines, were calculated to make a deep impression on the Indians. But not even a glance did the latter, as they approached, bestow on the military array. They seated themselves in solemn silence, and according to etiquette spent the next few minutes in smoking their pipes. A chief rejoicing in the name of Turtle-heart then made a pacific oration. Colonel Bouquet waited three days before making a reply. He assumed a high tone, reproaching the Indians with their cruelties and perfidy, and the violation of the engagements they had so recently made with Colonel Bradstreet. He then informed them that they could only escape condign chastisement by giving up within twelve days every single European prisoner in their hands and every negro. 'When,' added he, 'you have fully complied with these conditions, you shall then know on what terms you may obtain the peace you sue for.' The stern voice and resolute mien of Colonel Bouquet produced an immediate effect. Here was a man upon whom their native guile would be wasted, and their eloquence thrown away. The Red-skins were in fact reduced to a state of abject terror. They hastened to break up the conference and to fulfil the stipulated conditions. The Colonel, on his side, was determined to omit nothing that might confirm the effect of his speech. He marched thirty-three miles further down the Muskingum as far as Old Wyandot-town, in the heart of the Indian country, and halted to receive the prisoners at a spot where he had it in his power to destroy almost the whole of the Indian villages in the event of the chiefs breaking faith with him. Here, with unexampled rapidity four redoubts were thrown up, trees were cut down and the ground cleared. A town, furnished with store houses and hospitals, sprang up in the midst of the forest; buildings were constructed to receive the prisoners, and every precaution taken to fortify the settlement.

Just at this moment came a letter from Colonel Bradstreet at Sandusky to say that he was on the point of returning to Niagara. Had it occurred a few days sooner, this desertion would have encouraged the whole body of Indians to fight, and might have had most disastrous results; as it was, the terror inspired by Bouquet was sufficient; nay, he was able to undo the consequences of Bradstreet's folly and neglect. That officer, encamped as he had been, in close proximity to the Wyandot villages at Sandusky, had done nothing for the release of the prisoners with which the villagers were crowded. Bouquet at once sent a message to say that they were to be surrendered. At that distance he was not in a position to enforce compliance, but by a mixture of threat and diplomacy to a great extent gained his end.

The prisoners were now brought in from every side. By November 9, 206 had arrived, of whom 125 were women and children. Beaver and Custaloga brought in all they had but twelve, but until that small remainder had actually come, Bouquet refused to have the slightest intercourse with them. At length he was satisfied of their sincerity, his demeanour relaxed, and he offered peace, though even then he insisted on hostages, and referred them to Sir William Johnson for the precise conditions upon which it would be granted.

'It was a strange and moving sight,' says Parkman, 'when troop after troop of prisoners arrived in succession. The meeting of husbands with wives, and fathers with children, the reunion of broken families long separated in a disastrous captivity; and, on the other hand, the agonies of those who learned tidings of death and horror, or groaned under the torture of protracted suspense.'

But the children had, as a rule, been well treated by their Indian foster-parents, and there were some who had lost all recollection of father and mother; there were girls, too, now grown up, wedded to Indian husbands. It was with the greatest reluctance and in some cases under actual compulsion that they quitted the Red-skins, and rejoined their own race. One woman recognised her daughter who had been carried off nine years before, but the girl had entirely forgotten her mother and utterly failed to return her passionate embraces. Bouquet was standing by, and his suggestion at the moment reveals his imaginative mind. 'Sing her the songs,' said he, 'that you used to sing to your child.' Then memory returned, and with a flood of tears the girl buried herself in her mother's arms.

The Indians themselves were not less moved at the separation; and many sought and obtained leave to accompany the army back to Fort Pitt, hunting and bringing provisions *en route.* On November 28 the little army reached Fort Pitt on its return journey. With a generous and tactful word of praise for their steadiness and hardihood, Bouquet dismissed the levies of Pennsylvania and Virginia. He was splendidly received throughout the country, and at their ensuing session the Houses of both provinces assembled, passed a cordial vote of thanks to the able and intrepid commander. After detailing his services, the speaker of the Pennsylvanian Chamber went on as follows :

'These eminent services and such constant attention to the civil rights of His Majesty's subjects in this province demand, Sir, the grateful tribute of thanks from all good men ; and therefore we, the representatives of the freemen of Pennsylvania, unanimously, for ourselves and in behalf of the people of this province, do return you our most sincere and hearty thanks for these your great services ; wishing you a safe and pleasant voyage to England, with a kind and gracious reception from His Majesty.'

The Virginian Speaker spoke in similar terms.[1]

But as it turned out, Bouquet never did return to England. Both Houses had recommended that he should receive the rank of Brigadier-General, but although three years ago he had been naturalised as a British subject, it seemed most unlikely that their wish would be gratified, and it was with equal surprise and pleasure that he received the official intimation of his appointment. Needless to say none were more delighted at the well-earned honour than his own brother officers of the Royal Americans. Captain George Etherington wrote a letter which must have made him very happy.

'Lancaster, P.A.
'19*th April* 1765.

'SIR,—Though I almost despair of this reaching you before you sail for Europe, yet I cannot deny myself the pleasure of giving you joy in your promotion, and can with truth tell you that it gives great joy to all the gentlemen of the Battalion for two reasons :

[1] The praises of the House of Virginia are somewhat discounted by its refusal to pay the 200 volunteers whom Bouquet had induced to join him. In fact it threw the burden on his shoulders. The Assembly of Pennsylvania to its credit came to his rescue and assumed the debt itself. The matter caused the most intense mortification to Colonel Bouquet.

first on your account, and secondly on our own, as by that means we may hope for the pleasure of continuing under your command.

'You can hardly imagine how this place rings with the news of your promotion, for the townsmen and Boors stop us in the streets to ask if it is true that the King has made Colonel Bouquet a General; and when they are told it is true they march off with great joy; so you see the old proverb wrong for once which says "he that prospers is envied," for sure I am that all the people here are more pleased with the news of your promotion than they would be if the Government would take off the stamp duty.'[1]

Immediately afterwards General Bouquet was appointed to the command of all the troops in the southern colonies of British America. He sailed for Pensacola, his headquarters, soon afterwards. He was delayed *en route* for some months. We do not know the reason of the delay, it seems a little mysterious. He did not land until the third week in August. Within ten days he was no more.

Fever was no doubt the cause of death, but tradition has woven a wreath of romance over his grave and says that he died of a broken heart. Bouquet was only forty-seven. Of all his friends none mourned his loss more than his old comrade Colonel Haldimand. All that remained for Haldimand was to send bricks for a monument to his friend. '*Pulveris exigui munus.*' The memorial was soon washed away by the sea, but to those who would ask why it has not been replaced, we would point to the Royal American Regiment of that time and to the King's Royal Rifle Corps of our own, and in the words of Wren's epitaph reply, '*si monumentum requiris circumspice.*'

Far in advance of his age as his writings and career show Bouquet to have been, in England his name has

[1] The Stamp Act, which was extremely unpopular.

long been forgotten; in the American continent it is ever fresh, and to the present day, his spirit has dominated our Regiment, and will it is to be hoped continue to do so until time shall be no more.

Speaking of his campaign of 1763, and more particularly of his fight at Bushy Run, the Canadian historian Kingsford observes that—

> 'it may be described as one of the most prominent events of the continent, from the important consequences which followed. There has rarely been an action where gallantry, judgment, and determination could more justly claim public recognition. . . . But amid the dreary political contests of the home government it passed unrewarded. . . . His name remained without prefix to show the contemporary estimate of his worth; but whoever examines and considers his career by the results he achieved, the nobility of his nature, the absence of littleness of spirit, his sense of duty, his unfailing fortitude in the darkest hour, will say to all the world "this was a man." [1]
>
> 'Bouquet's leadership presents as brilliant a chapter in the history of the continent as may be found in its annals.'

The almost startling question forces itself upon us, had Bouquet lived, should we have lost America?

---

[1] When we recall the tradition of the romance which so deeply affected Bouquet's life; when we think of his hopes and ambitions unfulfilled, and of the irreparable loss inflicted by his death on the country of his adoption at a moment when his services were most required, we can hardly fail to recall Lewis Morris's lines as being not less appropriate to the mouth of our hero than to Endymion in the unseen world:

> 'Yet I judge it better indeed
> To seek in life, as now I know I sought,
> Some fair impossible Love, which slays our life,
> Some fair ideal raised too high for man
>
> . . . . . .
>
> Than to decline, as they do who have found
> Broad-paunched content and weal and happiness:
> And so an end. For one day, as I know,
> The high aim unfulfilled fulfils itself;
> The deep, unsatisfied thirst is satisfied.'

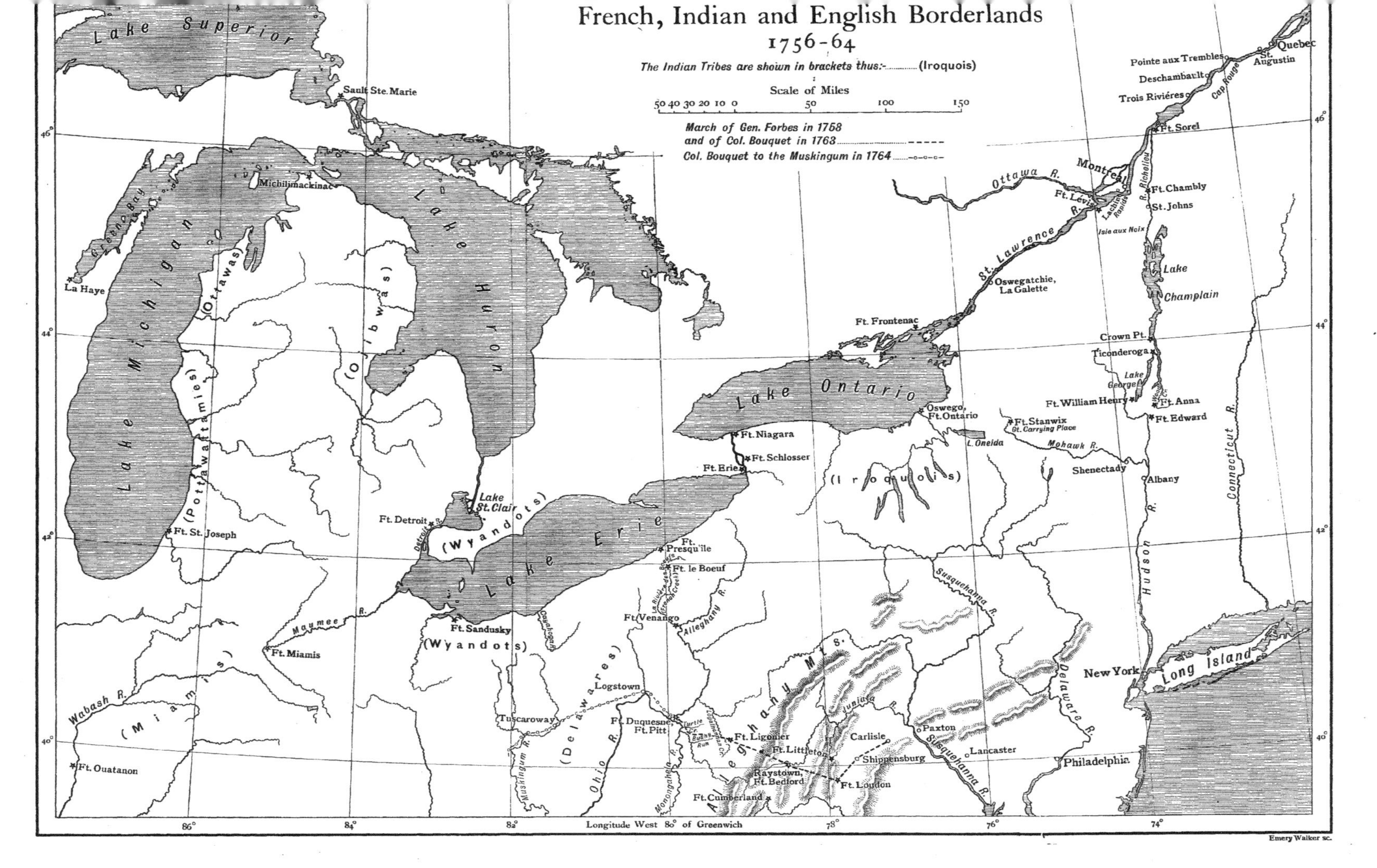
French, Indian and English Borderlands
1756-64
The Indian Tribes are shown in brackets thus:- (Iroquois)
Scale of Miles
50 40 30 20 10 0 50 100 150
March of Gen. Forbes in 1758
and of Col. Bouquet in 1763
Col. Bouquet to the Muskingum in 1764
Lake Superior
Sault Ste. Marie
Michilimackinac
Green Bay
La Haye
Lake Michigan
(Ottawas)
(Pottawattamies)
Ft. St. Joseph
(Ojibwas)
Lake Huron
Lake St. Clair
Ft. Detroit
Detroit R.
(Wyandots)
Lake Erie
Ft. Sandusky
Maumee R.
Ft. Miamis
(Miamis)
Wabash R.
Ft. Ouatanon
Cuyahoga R.
Tuscaroway
Muskingum R.
(Delawares)
Logstown
Ohio R.
Ft. Duquesne, Ft. Pitt
Monongahela R.
Ft. Cumberland
Ft. Ligonier
Ft. Littleton
Raystown, Ft. Bedford
Ft. Loudon
Shippensburg
Carlisle
Juniata R.
Allegheny Mts.
Paxton
Lancaster
Susquehanna R.
Philadelphia
Delaware R.
New York
Long Island
Hudson R.
Connecticut R.
Albany
Shenectady
Mohawk R.
Ft. Stanwix
Gt. Carrying Place
L. Oneida
Oswego, Ft. Ontario
(Iroquois)
Lake Ontario
Ft. Niagara
Ft. Schlosser
Ft. Erie
Ft. Presqu'ile
Ft. le Boeuf
Ft. Venango
Allegheny R.
Ft. Frontenac
Oswegatchie, La Galette
St. Lawrence R.
Ottawa R.
Ft. Lévis
Montreal
Lachine Rapids
Isle aux Noix
Richelieu R.
Ft. Chambly
St. Johns
Lake Champlain
Crown Pt.
Ticonderoga
Lake George
Ft. William Henry
Ft. Anna
Ft. Edward
Ft. Sorel
Trois Rivières
Deschambault
Pointe aux Trembles
Cap Rouge
St. Augustin
Quebec
Longitude West 80° of Greenwich
86° 84° 82° 78° 76° 74°
46° 44° 42° 40°
Emery Walker sc.

CHAPTER IX

THE COMING STORM—THE INTERLUDE—THE CARIB INSURRECTION—THE AMERICAN WAR OF INDEPENDENCE—THE THIRD AND FOURTH BATTALIONS RE-RAISED—THE SOUTHERN STATES—OPERATIONS OF AUGUSTIN PREVOST—BRILLIANT ACTION AT BRIAR CREEK—SUCCESSFUL DEFENCE OF SAVANNAH—LOSS OF ST. VINCENT—LOSS OF WEST FLORIDA—THE FIRST BATTALION AT NICARAGUA—THE UNITED STATES OF AMERICA

THE last line of Captain Etherington's letter, quoted in the preceding chapter, not only bears words of ill omen in its allusion to the hated Stamp Act,[1] but serves to show how very short was the interval between the cession of Canada and the revolt of our colonies. As fate willed, it was through the action of two officers of the Royal Americans that the strongest links binding the colonies to the mother country were broken. At Montreal Amherst had destroyed the French power on the continent, and at the Muskingum Bouquet had for ever removed the danger of the white man being exterminated by the red. With the disappearance of these two perils the turbulent spirits among the colonists could, without fear of consequences, indulge their love of bickering with the mother country; and the attentive listener could already hear the growling of the storm destined to burst within ten short years. But we anticipate.

[1] Under an Act of 1765 Stamp Duties were imposed on the American Colonies by the British Government. The Act was repealed in the following year.

Colonel Bouquet was succeeded in command of the 1st Battalion by his countryman Augustin Prevost, whom we last saw commanding the 3rd Battalion at Havana. Three companies had been sent to Carolina in 1764, the remaining six being now stationed at New York.

We left the 2nd Battalion at Quebec, where Major-General James Murray, our Colonel-Commandant, held command. The 15th and 27th Regiments were also included in the garrison where, on September 18, a serious episode occurred which shall be described by Bradley.[1]

'While George the Third and his Government were squandering thousands on unworthy legislators and sycophants, and in an unworthy cause, they selected this moment to exercise economy in the matter of their soldiers' wages. The mute heroes of many campaigns, who had driven the French from America and fought Indians through mosquito-haunted swamps in broiling summer days, were to be made to pay for their rations, hitherto free, at the rate of fourpence a day. Moved with indignation, the troops in Quebec assembled *en masse*, but without arms, before the Chateau St. Louis, and made complaint to the Governor. Some civilians, who upbraided the men, were pelted with stones, which fetched out the officers with drawn swords. Upon this the men ran to the barracks, seized their arms, formed in order, and, with beat of drum, moved towards the St. John's Gate. Murray himself now went among them, but to his appeals they replied that they would march to Amherst, the Commander-in-Chief, at New York, and lay their arms at his feet. They spoke with pride and respect to their officers, nor was anyone drunk, but the excitement was intense. The town-major, however, managed to close the gates, which created a panic lest the troops should mutineer and loot the city. Murray now persuaded them to march to the adjoining parade-ground, and earnestly besought them to remember their cloth, and return quietly to barracks. Addressing the officers, he dwelt on the certainty of a mutiny in Quebec spreading to the other garrisons, if successful, and the catastrophe thereby involved. Ordering a general parade, he again urged the men to obedience. They replied in praise of their officers, but resolutely refused to pay for their food. The night passed quietly, and on the next morning, Septem-

[1] *The Making of Canada.*

ber 20th, Murray told his officers that they must compel obedience to the obnoxious order, or die in the attempt; and the day was spent by them in fruitless attempts to talk their men round. On the next day a general parade was ordered, and the matter had to be put to the test. Murray, after reminding the men again of the "enormity of their crime," declared his fixed resolution, and that of his faithful officers, to compel them to submission, or perish in the attempt. He then went to the head of Amherst's grenadiers, he writes, "determined to put to death the first man who should disobey. Thank God I was not reduced to that horrid necessity." The whole company, followed by the entire body of troops, submitted, and marched quietly between the royal standards placed for the purpose, and back to the barracks.'

It is to be observed that Sir Jeffery Amherst was Colonel of two of the three regiments concerned, and, however much we may disapprove of the method adopted, it is a fair inference to suppose that the men intended only to appeal to their Colonel for redress of an acute grievance.

A year later the 2nd Battalion was moved to Montreal under somewhat curious circumstances. It happened that a regiment quartered there was in a high state of ferment on account of the cruelty of Mr. Walker, a magistrate. A fire had broken out with such violence that upwards of a hundred houses were burned down. It was winter. The wives and children of the soldiers were driven out into the snow without shelter, and Walker refused to quarter them upon the civilian inhabitants. It was said that some of these wretched people died of cold, and that certain officers and men of the regiment, disguised as Indians, thereupon entered Walker's house and cut off his ear.[1] The authorities thought it advisable to call in the services of a corps which was not in the habit of cutting off magistrates' ears; and our 2nd Battalion was accordingly brought up to Montreal in the depth of

[1] Six persons—some among them civilians, including a judge—were tried for complicity in this affair. All were acquitted.

winter, 1764–5. In the spring of 1765 it marched thence to New York, and a detachment was shortly afterwards detailed to protect a vessel containing stamps. The men were pelted by the mob; the excitement increased, but no magistrate appeared to read the Riot Act. Matters at length became so serious that the grenadier company charged and dispersed the crowd, but without bloodshed. That they felt indignant at the conduct of the citizens was shown a few days later when, on the point of marching for Albany, they broke open the prison, and by way of retaliation, liberated the Governor of Michilimackinac and several prisoners who had been confined for debt. Among the officers of our Regiment—and belonging, as it would appear, to this battalion and therefore possibly present at the time—was Major Horatio Gates,[1] whose fate it was twelve years later to receive, in the capacity of an American General, the surrender of a British army at Saratoga.

After a time the battalion, now once more commanded by Colonel Haldimand, gradually took the place of the 1st in the outlying solitudes. During the summer and autumn of 1765, three of its companies were stationed at Crown Point, three at Fort William Augustus, one at Fort Oswegatchie, one at Ticonderoga, and the ninth at New York. In the following year the battalion was even more split up, four companies being at La Prairie, two at Fort Erie, and the remainder at Fort Ontario, Detroit, and New York respectively. In 1767 two companies were stationed at Niagara, three at Detroit, two at Michilimackinac, one at Erie, and one at Ontario. With little change, the battalion, whose strength at the outset

[1] Horatio Gates, born in Essex; badly wounded with Braddock in 1755; A.D.C. to Monckton at Martinique: belonged to 45th Regiment; promoted to majority in 60th, October 1764. On quitting the army, settled down in Virginia. On outbreak of War of Independence offered his services to Congress; became successively Adjutant-General, Major-General, and Commander-in-Chief of the Northern Army.

little exceeded, and afterwards fell below, four hundred, remained in these wild quarters till 1772.

The 3rd Battalion had taken possession of West Florida after the peace of 1763, and early in the following year had been sent, *via* New York, to England to be disbanded; the 4th having preceded it a year earlier for the same purpose. The 1st and 2nd were each reduced to nine companies.

About this time the British Parliament passed an Act which had the most far-reaching consequences to the Royal Americans. In defiance of geographical considerations, in defiance of the actual and implied conditions under which the Regiment had been raised, it decreed that the West Indian Islands formed part of North America! Although the Act of Parliament under which the Regiment was raised used the word 'America' only, officers and men felt that the continent of North America had been implied, and considered that a breach of faith had been committed; but although (apparently at the representation of General Gage), Parliament, to gild the pill, gave to foreigners who had served in the Regiment for three years and were not members of the Church of Rome the privilege of exercising any employment, whether civil or military, in any part of the British dominions with the exception of Great Britain and Ireland, the obnoxious Act was not repealed, and in pursuance of it both battalions within the next few years were shipped off to the West Indies.

Five companies of the 1st Battalion were stationed at Quebec in July 1767. Thence, with an additional company, they sailed for Jamaica, and either in the same or the following year were joined by the three companies which had been on detachment in South Carolina. In total disregard of sanitary conditions the barracks were then situated on the low ground near Spanish Town. Malarial and yellow fever were prevalent; and in fact at that time the order for a regiment to proceed to those islands was equivalent to a sentence of lingering death. Such very nearly proved to be the fate of the 60th. The Regiment for almost half a century to come was destined to be quartered, except for comparatively short intervals, in the West Indies. The results were deplorable, and at the end of the period, the old Royal American Regiment had all but disappeared.

We have seen that as far back as 1759 a light company had been provisionally formed in our Regiment, but in the year 1770 a new departure was made throughout the army by the formation of a light company in each Battalion of the Line.

We left the 2nd Battalion split up in detachments at Niagara and the Lakes. In 1772, being then less than four hundred strong, the battalion embarked for Antigua, whence in January of the following year six companies were despatched to St. Vincent, where military operations under Brigadier-General Dalrymple were being conducted against the black Caribs—a hybrid race, the descendants of the aboriginal yellow Caribs and escaped African slaves. They were brave, fierce, and intractable; and in a mountainous country, destitute of roads and covered with thick forest, the task of reducing them to submission was difficult. On the arrival of our 2nd Battalion, Dalrymple found himself strong enough to drive the Caribs into the mountains, where they had only the alternative of death by famine or surrender. They quickly chose the latter. According to Major Patrick Murray, something of this happy result was due to the diplomatic talents of George Etherington, by this time a Major. Murray states that Etherington marched the light company to the Carib town on the sea-shore, and 'being a man of consummate address among the lower orders of mankind, brought them to capitulation.' His good offices did not remain unrewarded, for the grateful Caribs granted him a valuable tract of land. The loss of the troops in action was small, but the climate made ravages in their ranks, and at least twenty-five per cent. of the whole force was on the sick list.

Six companies of the battalion remained at St. Vincent; the four companies left at Antigua stayed there under command of Brevet-Major Brown.

And now matters were fast tending to 'that great tragedy of errors,' the revolt of the colonies, or, as our Transatlantic cousins naturally term it, the American War of Independence. The lapse of time has softened the asperities engendered by that unhappy contest. We are able to delineate its episodes with truer perspective, and to congratulate the citizens of the United States on the progress since made, and the high degree of prosperity attained. Yet even so, it is not quite easy for any but one of American birth to wax enthusiastic over the circumstances which led to separation. Far be it from us to do less than justice to the nobility of impulse which decided that at whatever cost old traditions must be broken, and that true allegiance was due rather to the land of their birth or settlement than to her whom they had previously regarded as their mother country. It must be remembered that the Pilgrim Fathers, when they landed on the shores of the New World, had—to use George Canning's words of a later date—made their perilous voyage with the express purpose of 'bringing a new World into existence to redress the balance of the Old'—

'Teucer Salamina patremque
Cum fugeret . . .
Sic tristes affatus amicos
"Quo nos cumque feret melior fortuna parente
Ibimus . . .
Certus enim promisit Apollo
Ambiguam tellure nova Salamina futuram."'

Nor do we wish to decry the more recent ideals of freedom and liberty, even if in our own minds we question whether they were not somewhat thin and one-sided. The project of independence was no new one in America. It had been mooted so far back as 1689, and force alone had for the time being compelled its abandonment. But the aspiration had never been entirely relinquished, and it was so plain to Vaudreuil and others, that as far back as 1763 they had said in so many words that the cession of Canada would be the inevitable prelude to the independence of the American colonies.

Nevertheless, however grand the ideals which may have pervaded the hearts of many persons, however noble their aspirations, it can hardly be denied that there was much in the conduct of their rulers which grates on one, even at this distance of time. That men now shouting for independence should but a few years previously have refused to raise a finger to protect their own people against Indian savagery strikes one as being somewhat anomalous; while the fact that Boston magistrates had recently sold into slavery an unfortunate British private soldier unable to pay an overwhelming fine which they had inflicted on him, seems to show that the love of liberty so loudly boasted was, at all events on occasions, of a somewhat partial character.

The inhabitants of New England for the most part came of a Puritan and Roundhead stock, and could hardly be expected to have much affection towards the monarchy of Britain. In the southern states the case was rather different. A proportion of the inhabitants was of cavalier descent and disposed to cherish the connection with England. But the paramount importance of the Sovereign being personally known to the colonies was not at that time understood at home; while the barbarities with which the 'Forty-five' had been suppressed taught far and wide the indelible lesson that fidelity to an absent dynasty was the most unpardonable of crimes.

'The actual point at issue,' writes the Hon. John Fortescue, 'was the sharing of the burden of Imperial Defence; but this was in reality but a particular phase in a dispute which was nearly as old as the colonies themselves. It was the old notion in Europe that no colonies could be of profit to a mother country unless she monopolised their trade. Accordingly a succession of Acts, of which the principal dated from 1660, had been passed to ensure that no colonial produce should be imported into England except from English colonies, and no European produce into English colonies except by way of England. These enactments had given trouble from the very beginning, having for years been openly defied in New England, and afterwards evaded by a general system of smuggling in all the colonies. The result was that the revenue gathered under the Acts of Trade in the colonies, which by right should have been very considerable, in actual fact did not cover the cost of collection. At the close of the Seven Years' War, when England was staggering under the enormous cost of the late operations, it occurred to George Grenville, the Prime Minister, that the Exchequer would profit considerably if these Acts were rigidly enforced. Moreover, since the conquest of Canada had been undertaken mainly for the benefit of the American colonies, it was right that they should contribute something towards the cost of maintaining the necessary garrisons in the conquered territory.

'It was therefore proposed that 10,000 men should be quartered in America, and that £100,000 a year should be raised from the Colonists for their maintenance. Grenville would willingly have accepted a voluntary contribution to that amount from the various colonies; but past experience had shown that the thirteen provinces could never agree as to the proportion of any common burden that should be borne by each of them; and he therefore sought to save undue trouble and friction by collecting the sum through levying Stamp-duties under an Act of the British Parliament. This brought up the question whether the colonies could be taxed by an Assembly in which they were unrepresented; and the American leaders, very cleverly and adroitly, seized the opportunity to conceal the true issue—namely, whether the colonies should contribute to the cost of Imperial Defence—and to concentrate all attention upon the constitutional principle that there could be no taxation without representation. This policy served a double purpose. First it diverted attention from the shameful apathy with which the colonists had allowed the Imperial troops to bear the brunt of the war against Pontiac; next it gave them strong ground for refusing obedience to the detested Acts of Trade. The British Ministers hastened to repeal the Stamp Act; but to preserve the integrity of the Acts of Trade they maintained the right of the British Parliament to tax the colonies as persistently as the colonies impugned it. The dispute waxed more and more bitter; and the mob of Boston supported the colonial contention by repeated acts of violence against British soldiers and British officials, acts which the executive of Massachusetts was too weak to suppress or to punish. Meanwhile the British Government took up an attitude which was at once weak and oppressive, strongly asserting the right of the Imperial Parliament to tax America, but furnishing no sufficient force to the King's Governors to keep the mob in order. Matters therefore drifted on until the Americans, emboldened by success, resolved to organise armed resistance and to stand or fall by the chances of war. The British forces in the American colonies were few, and could not be readily reinforced. Every advantage, therefore, was to be gained by them if they took the initiative, and there were plenty of men among them wise and shrewd enough to be aware of it. On the other hand the British Ministers, long after all chance of a peaceful settlement had passed away, clung desperately to the hope that war might yet be averted. Thus there was long wavering between coercion and conciliation, with the natural result that no efficient force was sent at the outset of the quarrel.'

Blood was first shed on April 19, 1775, when a small force of British soldiers despatched by General Gage to seize a magazine, encountered American militia drawn up to oppose them, and during its return after the execution of its object, was set upon for many miles by the armed population of the villages *en route*, and made good its retreat only after a loss of 260 men. This was called the affair of Lexington. Following up their success, 20,000 Americans blockaded eleven weak British battalions in Boston; and although General Gage gained a pyrrhic victory at Bunker Hill, the situation was so little improved that in the following March General Howe, whom we last met in command of Wolfe's light troops on the Plains of Abraham, and who succeeded Gage as Commander-in-Chief, found himself obliged to evacuate the town, and sail for Halifax. On the other hand, an American force which had surprised Ticonderoga and invaded Canada, was completely defeated and driven out of Canada by General Carleton, formerly Wolfe's Quartermaster-General. So far, things were about even; but now appears again on the scene our old acquaintance George Washington, whom a wise decree of Congress had appointed Commander-in-Chief of the American army at Boston. His noble figure towers above those of his compatriots, and if ever a man deserved well of his country it was Washington.[1]

Events moved apace, and on July 4, 1776, the Congress of the thirteen colonies declared the independence of the United States of America. It was a bold course, for the tide of their fortunes was running low. Carleton advanced victoriously to Lake Champlain, but unfortunately stopped short of Ticonderoga. Howe captured New York, and Washington found his army melt away to a strength of 3000 men. But in the winter, the latter, resuming the offensive, forced the British to evacuate almost the whole of New Jersey.

At this crisis the Americans found an invaluable ally

[1] Washington's fortitude and constancy deserve the highest praise. To us his weak point seems to be an absence of chivalrous feeling either towards British troops or American loyalists.

in the person of Lord George Germaine, the British Secretary of State. Germaine under his previous name of Sackville had been cashiered for misconduct at the battle of Minden; but was even more inefficient as a minister than as a soldier, and it is to his direct instructions that our disasters at this period can be traced. He superseded Carleton; and, having replaced him by General Burgoyne, ordered the latter to advance southwards upon Albany and to join Sir William Howe on the Hudson River; while at the same time he allowed Howe to undertake independent operations in Pennsylvania. The result was that in the summer of 1777 Sir William landed at Chesapeake Bay and defeated Washington on the Brandywine River, captured Philadelphia, and once more repulsed Washington on October 24 at Germantown; while Burgoyne, relying upon the promised co-operation, made his way southwards with great difficulty, only to find himself isolated; and, being surrounded by an overwhelming force under General Gates, was compelled on October 16 to surrender with the whole of his Army.

In February 1778 France concluded a treaty of alliance with the United States. It was impossible for British troops to occupy the whole of the colonies, and all the northern and middle states with the exception of the city of New York, were in consequence evacuated. But in the south—Virginia, the Carolinas and Georgia (whose population was upwards of a million—rather more than one-third of the whole)—great hopes were entertained on account of the existence of a large body of loyalists, and it was determined to concentrate all efforts on those provinces.

But to return to the Regiment. At the time of the outbreak of the war nearly all the old names had disappeared. Jacques Prevost, now Governor of Antigua, was, however, Colonel-Commandant of the 1st Battalion, and his brother Augustin was his lieutenant-colonel, while the other brother, Jean Marc, was a captain and brevet-major in the same battalion. Augustine Prevost was his father's adjutant, and in fact, during the next few years, the Royal Americans contained so many members, legitimate or otherwise, of the same family that it has been called the Prevost Regiment.[1] Bernard Ratzer and Des Barres still remained in the Regiment; so did Christopher Pauli, whose sleep even at this distance of time may have been disturbed by the nightmare of a suit in the Divorce Court by his Indian wife for restitution of conjugal rights. In the 2nd Battalion Frederick Haldimand, now major-general, was colonel-commandant; George Etherington was major, and among the captains we still find Daniel McAlpin, who had led the light company up to the Heights of Abraham, and Dietrich Brehm who had been sent by Rogers to summon the French commandant at Detroit. John Christie was still a subaltern in the battalion. Among other original members of the Royal American Regiment was John Muller, Christopher Spiesmacher, Ralph Phillips, James von Ingen, and Thomas Grandidier.

It seems obvious that from its training and associations the Royal American Regiment was highly fitted for the warfare now entered upon; but it seems an equally obvious deduction that although no want of loyalty had ever been shown, yet in thought, habit, and association it was so thoroughly American that it would have been unfair to put it into the field against the colonists. Indeed, so great a degree of sympathy with the latter was felt by naval and military officers whose duties had brought them to America that several, although they had long returned to England, refused to serve in the war at all, and among

[1] The Bontems and Bonteurs whose names we also find in the Regimental Lists were also related by marriage to the Prevosts.

these was Sir Jeffery Amherst. It was, therefore, not unnatural if, in the Royal Americans, opinion should be sharply divided. The instance of Gates is, perhaps, not a very good one, as his service in the Regiment was extremely short; but in the André tragedy we find one ex-officer of the 60th pleading for the major's life, while another, Arthur St. Clair, sat as an American general on the court-martial, and signed the death warrant.

We are consequently not surprised to find the 1st and 2nd Battalions left in the West Indies and taking little part in the war; but one of the first measures adopted by the British Government was to raise once more the 3rd and 4th Battalions, placing both, as well as the 1st and 2nd, on an establishment of 650 N.C.O.s and men. Two hundred men were enlisted by Augustin Prevost in the outskirts of Hanover, others in England, whence they embarked for Florida.[1] Headquarters and three companies of the 3rd Battalion under Lieut.-Colonel Stiell[2] were despatched to Pensacola. The remaining companies under Major Van Braam, and six companies of the 4th Battalion were quartered at St. Augustin, the other companies of the 4th Battalion being sent to West Florida. Augustin Prevost, by this time Colonel-Commandant of the 4th Battalion, took command of the force in East Florida, and was joined early in 1777 by Lieut.-Colonel J. M. Prevost with three weak companies of the 2nd Battalion which

[1] The Muster Rolls give the following figures, showing that the numbers actually enlisted were far below the establishment:

Nov. 24, 1776.	Companies		Present	Absent	Casuals (i.e. belonging to the Companies detached)	Total.
3rd Battalion	6	Officers	3	11	2	16
		N.C.O.s and Men	112	21	154	287
4th Battalion	6	Officers	12	9	—	21
		N.C.O.s and Men	187	33	76	296

[2] Stiell, like Des Barres, lived to be well over 100. He died in 1810.

Amherst had directed to be sent from St. Vincent with a view to training the newly formed battalions in the methods and spirit of the Regiment. George Etherington, now Lieutenant-Colonel of the 2nd Battalion, seems to have obeyed the order unwillingly, for we are told that the three companies detached consisted only of Lieutenants Wickham and Muller with 82 N.C.O.s and men. Lieut.-Colonel Fuser with Captains M'Intosh and Murray arrived, however, from the 1st Battalion; M'Intosh being posted to the 3rd Battalion, the other two to the 4th. Other veteran officers of the 60th also joined the 4th Battalion, and among its subalterns we now for the first time find the name of Harry Burrard, thirty years later a war-worn veteran of repute, yet to us of the present day known chiefly from the fact that he superseded Sir Arthur Wellesley during the battle of Vimiero.

For a detailed account of the field operations of the Regiment during the early months of 1778 the reader is referred to Major Murray's memoir, *vide* Appendix IV; it must suffice here to give a summary of its movements.

In conjunction with the loyalists of Florida a guerilla warfare was carried on with success against the Militia of Georgia and South Carolina under Major-General Robert Howe. During the autumn Howe was placed under the orders of General Lincoln.

Meanwhile Sir H. Clinton, the British Commander-in-Chief, had been ordered from home to occupy Georgia and advance thence into South Carolina. To this end a force of 3500 men under Lieut.-Colonel Archibald Campbell of the 71st was despatched in November by sea to the mouth of the Savannah River, where it arrived on December 23. Lincoln had assumed command at Charleston four days previously, but found there only

1500 men. Howe, who with the 1st and 2nd Georgia and the 3rd and 4th South Carolina Regiments was posted at Fort Sunbury, 20 miles south of Savannah city, hearing however of Campbell's arrival, left a garrison in the fort, and marched with the bulk of his force to defend the town.

During the previous month Colonel Augustin Prevost had received orders to march with all available troops upon Savannah; there to form a junction with Campbell. In pursuance of these orders, Prevost quitted St. Augustin with a force of about 900 officers and men, of whom 20 officers and 400 men belonged to his own regiment;[1] the remainder being composed of a detachment of the 16th, some provincial corps, and a few artillerymen. His brother accompanied him as second-in-command, while Major Beamsley Glazier of the 4th Battalion was in command of the Royal Americans. The problem of transport was extremely difficult: supplies had to be convoyed in boats along the coast; and as the boats were often compelled to avoid the armed vessels of the Americans and make a wide detour, they were frequently separated by long distances from the troops whom they were intended to feed. The march of the column was therefore difficult and painful, and for some time the men actually lived upon the oysters found in inlets of the sea, and a very small supply of rice; in fact for three days the force subsisted on the rather remarkable diet of an alligator and some Madeira wine which had happily been saved from a wreck. But Prevost was not to be denied: he got near Sunbury in Georgia, and sent forward his brother to surround it. The American Commandant having refused to surrender, trenches were opened and guns brought up.

[1]

2nd Battalion,	5	Officers,	77	N.C.O.s and men.
3rd ,,	4	,,	100	,, ,, ,,
4th ,,	11	,,	223	,, ,, ,,

AUGUSTIN PREVOST

On January 9, 1779, the fort capitulated:[1] 212 prisoners and 45 guns were captured, and Prevost marched twenty miles northward to Savannah which had already been captured by Colonel Campbell. He formed a junction with that officer on the 17th, and assumed command of the combined forces, amounting apparently to between 4000 and 5000 men. Lieut.-Colonel Fuser of the 60th with 250 men of the 4th Battalion co-operated with Prevost's movement, forming a guard for his outer flank.

Augusta, the second town of Georgia, on the Savannah River a hundred and twenty miles north-west of the capital, was occupied a few days afterwards, but almost immediately abandoned; for the American General Lincoln, who had been reinforced to a strength of nearly 4000 men, had reached Purysburg, only fifteen miles above Savannah—a movement which forced Prevost to concentrate at Ebenezer, on the right bank of the river. Colonel Campbell, who had been in command of the force detached to Augusta, retired slowly down the Savannah River to Hudson's Ferry, twenty-four miles above Ebenezer, where, meeting Marc Prevost, he handed over his column to that officer. Meanwhile, Lincoln's force had been continuously reinforced to 7000 men; and ere long he found himself strong enough to detach General Ashe to occupy Augusta with some 2000 men. Ashe, finding that the British had withdrawn, marched down the river to its junction with the Briar Creek fifty-three miles above Savannah, where, on February 27, he encamped at a point but thirteen miles above Hudson's Ferry. His flanks were protected by the two rivers. His retreat was secured by a bridge, but he little knew with whom he had to deal.

Marc Prevost quickly located him; selected a detachment of about 900 men and two guns including the

[1] The 60th lost one man killed and three wounded.

grenadier companies of our 2nd, 3rd and 4th Battalions and a few other men of the 60th, with the 2nd Battalion of the 71st and Tause's Light Dragoons; described a wide circle with a radius of fifteen miles from Briar Creek, and established himself in Ashe's rear. On March 3, the 71st with two guns held Ashe in front; when, to the amazement of that General, Prevost at the head of the 60th burst like a thunderbolt upon his rear. The surprise was complete: Prevost's loss was only 16 men killed and wounded. Of the enemy, 150 fell in the action; Brigadier-General Elbert, 26 officers, and over 200 men were made prisoners; 150 others were drowned in the attempt to escape across the river, while seven guns with the whole of their ammunition and their baggage were captured. So complete indeed was the rout of Ashe's force that out of over 1500 men engaged not 500 ever rejoined General Lincoln. This was, it is true, an operation of minor tactics but cleverly executed; highly creditable to the skill and enterprise of Marc Prevost, and speaking volumes for the efficiency of the Royal Americans as light troops. 'This brilliant action of the British,' says the distinguished American general and writer, F. V. Greene, 'destroyed the possibility of recovering Georgia at that time.'

The state had been entirely recovered for the King, but General Lincoln received still further reinforcements. Leaving a detachment to garrison Purysburg and Black Swamp, he marched once more upon Augusta at the head of 4000 men. Augustin Prevost promptly countered by invading South Carolina with 2000 men, and finding he could not entice Lincoln to return, resolved to make a dash upon Charleston some eighty miles north of Savannah. A sudden rise of the water saved the American garrison at Black Swamp; but on marched Prevost in spite of the fact that the bridges over innumerable creeks had been destroyed. On May 12 he appeared before Charleston; and having summoned it to surrender, so far intimidated the local authorities that they proposed the neutrality of South Carolina during the war. Its garrison was superior in numbers to the British force; its defences were strong; nevertheless Prevost would not listen to the proposal. But by this time Lincoln was rapidly approaching from Augusta. The British force was in danger of being crushed between the upper and the nether mill-stone, for the enemy was in possession of the Savannah road. Finding his coup had failed Prevost—by this time promoted to the rank of Major-General—retired across the Ashley River and occupied comfortable quarters in James Island and John's. Here supplies were abundant, and the sea afforded him a retreat in case of need. His little campaign had been a notable one, and many would have been satisfied with its performance. His own modest view was far otherwise. 'I am in bad health,' he writes, 'and too infirm for the Command; had I possessed my old activity when pursuing the rebels through Carolina I should have been at Charleston as soon as they, and have captured it without firing a shot.'

The peril of Charleston had, as we have seen, brought Lincoln back in hot haste, and on June 4 he established himself opposite the British posts in hopes at all events of capturing some. But his attack was repulsed. Prevost, leaving a garrison on Port Royal Island, then withdrew his troops slowly from island to island until he regained Savannah. The heat became intense and for a time operations languished; but Captain W. Wolff was killed in an attack upon a privateer, and Captain J. Muller in a skirmish on the Ogeechee, whose fords he had been posted to defend but imprudently crossed.

But now we had lost command of the sea: a French fleet under Admiral Count D'Estaing[1] had already achieved success in the Caribbean, and the rebel Governor of South Carolina invited him to the coast of Georgia. D'Estaing at once complied; and on September 4 appeared before Savannah with 20 ships of the line and 11 smaller vessels carrying 6000 soldiers and over 2000 guns. So complete a surprise was his arrival that two British ships were taken at anchor. Augustin Prevost, who had received no warning, at once called in all outlying detachments, and set several hundred negroes to work upon the defence of the place. On the 10th the French fleet entered the river, and on the 15th D'Estaing having landed the greater part of his troops, summoned Prevost to surrender. The garrisons of Sunbury and Beaufort, on Port Royal Island, had not yet come in, and the British General, anxious to gain time, requested twenty-four hours to consider his answer. Just before the time had expired the force from Fort Royal arrived, raising the strength of the garrison to 3700 men, of whom, however, only 2000 were fit for duty. Among these were 200 of the 60th, mostly from the 4th Battalion, besides the grenadier companies and a few men included in a composite mounted infantry corps under Captain Tawse. Of the remaining troops about 1100 men belonged to the regular army; there were two battalions of Hessians and about half the entire force consisted of provincial levies.

Having concentrated his garrison, General Prevost replied to D'Estaing's summons in terms of defiance. On the north side, the town of Savannah was covered by the

[1] Count D'Estaing served under Lally in India when he was taken prisoner by the British. During the American war he served in the navy. At the opening of the French Revolution he was in command of the National Guard, but despite his popular opinions he was guillotined in 1793.

river; on the west by a morass; on the south and east the whole line of defence did not exceed three-quarters of a mile in length, and every yard had been fortified by abattis and entrenchments. At the south-eastern and south-western angles (the latter being termed Spring Hill) redoubts were thrown up, and three others occupied advantageous positions in the intervening space. By the irony of fate one contingent of the attacking force was commanded by an Englishman; the land force of the other by Count Dillon, a man of Irish descent, while the British defence was in the hands of a Swiss.

General Lincoln soon brought up his force to combine with D'Estaing, but the enemy's siege train was so much delayed by want of horses that it was not until the 23rd that the Allies were able to break ground. Prevost had however, in the meanwhile kept them well employed: twice he sallied forth, once with so much skill that he contrived to set French and Americans firing upon one another. On October 4 the besiegers opened fire from fourteen mortars and fifty-three heavy guns. The bombardment continued until the 8th without doing much damage. The French Admiral felt that time was precious, for none could say how soon the British fleet might appear in the offing, or how quickly autumn storms might drive him off the coast: he therefore resolved upon an immediate assault.

The allied forces consisted of about 10,000 men. General Huger with the Carolina Militia was directed to make a demonstration against the southern and eastern fronts: two columns of regular troops, including 3500 French, and 850 Americans under Count Pulaski, a Pole of distinction, received orders: one, under Count Dillon, to creep along the swamp on the western front and enter the defences from the rear; the other, commanded by

D'Estaing and Lincoln in person, to storm the Spring Hill redoubt.

Before dawn on the 9th the columns advanced to the assault. That appointed to pass the western front lost its way in the marsh, and, being unable to extricate itself, became exposed to the full fire of the British batteries at daylight. Eventually it retired without ever reaching its point of attack. The other column, well led, almost reached the Spring Hill redoubt without being discovered. Then the defending guns opened a terrible fire; but the Allies were not to be denied, and a hand-to-hand contest took place with the garrison of the redoubt, consisting of 100 men of our 4th Battalion and of the dismounted dragoons under Captain Tawse of the 71st. The fighting was desperate: Tawse cut down three men but, unable to extricate his sword from the body of the third, was himself killed.

So critical was the moment, so keen the assault, that a colour of the 2nd South Carolina Regiment commanded by Lieut.-Colonel Marion was planted on the rampart of the Spring Hill redoubt by Sergeant Macdonald, who had snatched it from the hand of Lieutenant Gray as the latter fell backwards mortally wounded. The other colour of the regiment was carried as far as the ditch when its bearer, Lieutenant Bush, received a fatal shot and dropped dead. After the action this colour was picked up under Bush's body by the Royal Americans and handed to their Colonel-Commandant, General Prevost, in the possession of whose great-grandson it still remains.[1] Macdonald's

[1] The 2nd South Carolina Regiment had had the post of honour in the successful defence of Fort Moultric, Charleston, against the British fleet, on June 28, 1776. A few days afterwards the wife of Major Barnard Elliott presented to the regiment a pair of colours—one blue and the other red—saying: 'Gentlemen, soldiers, your gallant behaviour in defence of your country entitles you to the highest honours. Accept these two standards as a reward justly due to your regiment: and I make not the least doubt that you will stand by them as long as they wave in the air of

Flag of the 2nd. South Carolina Regiment
Captured at Savannah

colour was safely carried off when the assailants retired, but fell into British hands at the subsequent capture of Charleston.

Meanwhile the fight showed no sign of flagging. Overwhelmed as it was by numbers—the odds seem to have been five or six to one—the garrison of the redoubt refused to give way and for nearly an hour the fight still hung in the balance. But a counterstroke was at hand. Huger having by this time been repulsed, reinforcements could be spared from the eastern side; and Major Beamsley Glazier bringing up the three grenadier companies of the 60th, with a detachment of Marines, charged the assailants with such fury that the French and Americans were driven back in headlong rout. As they fled they came under the fire of the British batteries; 900 dead or wounded men were left behind, 636 of whom were French and 264 Americans: 270 dead bodies were buried, 80 of which were lying in the ditch or upon the parapet, and 93 within the abattis—a proof of the determined valour displayed by the allies. This action, wherein Captain Wickham of our 2nd Battalion distinguished himself, had lasted two hours, during which the British casualties were only 16 killed and 39 wounded. Of the former, four, and of the latter, seven, belonged to the 60th. A dense fog prevailed, and Prevost decided not to pursue. But the lesson had been enough. The French and Americans quarrelled on the spot; the siege was raised; Count D'Estaing, who had been severely wounded, retired to his fleet

liberty.' At the assault on the Spring Hill redoubt, Lieutenant Bush being wounded handed the blue colour to Sergeant Jasper. Jasper who had already received a bullet was then mortally wounded, but returned the colour to Bush who the next minute fell, yet even in the moment of death attempted to protect the flag which was afterwards found underneath him. No one could have done more; and the colour (shown facing p. 216), hallowed by the blood of Bush and Jasper, deserves to be deposited under a consecrated roof.

and sailed away; while Lincoln returned to South Carolina.

The consequences of Prevost's brilliant defence were important. Georgia was cleared of the rebels, and the sanguine expectations throughout the United States that the reduction of the Province would lead to the abandonment of the colonies by the British, were for the time being disappointed. The historian Stedman remarks:

> 'Such was the termination of the siege of Savannah, during which it is said that the allied armies lost in killed, wounded and by desertion more than 1500 men, whereas the loss of the garrison did not exceed 120.
>
> The cool, steady, prudent and firm conduct of General Prevost . . . rank the successful defence of Savannah amongst the most brilliant achievements of the war.'

During this period the French had been busy among our West Indian possessions with varied success. In September 1778 they captured Dominica, but in December lost the more important island of St. Lucia. In June 1779 Count D'Estaing took advantage of the temporary absence of the British fleet to despatch a squadron with orders to get possession of the island of St. Vincent. This island was garrisoned by seven companies of the 2nd Battalion of the Royal Americans: the battalion was numerically weak, reckoning but 461 men, and of this small number upwards of 100 were on the sick list. Its principal duty was in reality to form a police force and hold the Caribs in check. To this end it had been scattered all over the island. On June 12 the French made their appearance, and having at once landed a party certainly not more, and probably less, than 450 strong, marched upon the capital and summoned the Governor. The latter, Mr. Valentine Morris, was at loggerheads with his

Council, a fact which perhaps impeded any measures he may have contemplated; anyhow he at once surrendered the island upon the condition that the Regiment should march out with the honours of war, should be sent to Antigua, and there exchanged for an equal number of French prisoners. The name of Colonel Etherington, commanding the 2nd Battalion, does not appear upon the instrument of capitulation; but that circumstance in itself by no means exonerates him. We do not know the exact facts of the case and are not in a position actually to assign censure to anyone or to assume that it was even possible to concentrate the garrison for defence. But be it remembered that no officer of standing divests himself of responsibility by acquiescence in the deed of a civil Governor, if such deed is detrimental to the military interests of the post which British troops have been sent to protect. If Etherington saw his way to defend the island, it was his business to do so whatever might be the views of the Governor. Nay more; each officer present is not only within his right, but in duty to his country is bound, to refuse to acquiesce in surrender arranged by his senior officer, if he believes it to be within his power to induce the troops to continue the defence. It is perfectly true that in acting thus he incurs a grave responsibility; for he renders himself liable to summary execution if subsequently captured by the enemy, and to the verdict of a court-martial if he return home in safety; yet even though he ultimately fail to defend his post more honour will accrue to an individual officer who thus loses his life or commission than to his superior who has surrendered the troops under his charge; and if he succeed immortal glory will be his reward. The practice of dispersing the British land force in driblets throughout the West Indian Islands was undoubtedly wrong; and it may well have happened that the Battalion of the 60th in the present instance was of more ultimate use elsewhere than it would have been in holding on to St. Vincent; but such considerations are not for the officer commanding a garrison. Rightly or wrongly he has been assigned a post, and it is his business to defend it to the very last. Remember Bouquet's words: 'If an officer should remain alone in his post, then he must die before he disgrace himself by abandoning it.'

We now return to Florida. During the autumn of 1778 it had become evident that Spain intended to join in the combination against Great Britain. The former kingdom still held possession of New Orleans and the surrounding tract of country; and orders were sent by the British Government to Colonel Campbell who commanded a weak force at Pensacola, to attack New Orleans as soon as war should be declared. Under these conditions Campbell had good reason to expect to receive the earliest news of that event; but on the contrary, Don Bernardo de Galvez, the Spanish Governor, received news of the declaration of war, which took place in June 1779, several weeks earlier than did Campbell. De Galvez at once marched with 2000 men against the British force which as usual was by the orders of Government disposed over many points, and too weak to defend any. Campbell's whole force seems to have comprised seven companies of the 16th Regiment—worn-out veterans—four companies of the 3rd, and the same number of the 4th Battalion of the 60th: together with a useless battalion of men from Weldeck, and two battalions of American Irish who had deserted the rebels and were not less ready to desert the British. Of money Campbell found himself destitute; he had neither entrenching tools nor boats to keep up communication with his outlying posts on the Mississippi. Pensacola was unhealthy; so was Mobile; so were the forts on the Mississippi, and his already scanty force was dispersed among them all. A fort had been built at the confluence of the Iberville River with the Mississippi, and it was this post which bore the brunt of de Galvez' attack. The Commandant, Lieut.-Colonel Dixon, had no knowledge of the Spanish declaration of war; but, realising at the last moment how matters stood, entrenched himself at Baton Rouge, and prepared to defend his position. His detachment consisted of about 500 men, including apparently three of our 3rd Battalion companies. For nine days it held out; then the Spaniards brought up heavy guns, and on September 24 Dixon capitulated. Among the killed was Lieutenant Ferdinand Brock of the 60th. In January 1780 de Galvez organised an expedition against Mobile, where the garrison was rather more than three hundred strong; but—except for four weak companies numbering 98 men of our 4th Battalion under Captain John Christie—composed entirely of provincial levies. The ramparts had been so much neglected that the fort seemed defenceless; but the Commandant, Captain Durnford of the Engineers, was determined to make the best of it. About the end of January de Galvez made his appearance, but with only a few of his transports and a single man-of-war; for the bulk of his fleet had been dispersed by a storm. All that was needed, said Campbell afterwards, was one British frigate; but no such frigate was there, and de Galvez having landed the troops that were with him awaited reinforcements. These gradually arrived; and having concentrated 1500 men the Spanish General on February 25 disembarked his troops a second time—on this occasion within four miles of Mobile. In spite of the miserable state of the defences the Spaniards hesitated to assault.

Siege operations were at once begun, and on March 14 the Spanish batteries opened fire. Colonel Campbell had lost no time in marching to the relief of the place, and arrived near enough to hear the report of the guns. On a sudden they ceased. The bombardment of a very few hours had made the wretched place untenable. Captain Durnford was unaware of Campbell's approach; he had no further means of defence, and had made the best terms in his power. The garrison marched out through the breach. Lieutenants Loup and McDonald of the 60th are said to have distinguished themselves in the defence, and Christie would appear to have recovered the reputation he had lost in 1763 at Presqu'île. The four companies of our Regiment were taken to Vera Cruz and Cuba, but rejoined the British Army at St. Augustin in 1781.

During all this time the 1st Battalion of the Royal Americans had been quartered in Jamaica, the Governor of which was Major-General John Dalling, a veteran of the Regiment and at the present time Colonel-Commandant of the 3rd Battalion. Dalling unfortunately conceived the idea of attacking the Spanish settlements in Nicaragua. His project was to despatch a small force to advance up the St. Juan River as far as Lake Nicaragua; pushing on thence to the Pacific Ocean and cutting Spanish America in two by a chain of posts from the Pacific to the Gulf of Mexico. He was assured that the Spanish colonists were only awaiting such a force to emancipate themselves from their mother country. British Ministers welcomed the idea with enthusiasm. The fallacy of initiating military operations which were to depend upon local support had just been shown in Carolina and Georgia: within twenty years it was destined to be once more exemplified in the bloody tragedy of Quiberon Bay; on the present occasion it was equally manifest. Captain Polson of the 60th was given the rank of Lieutenant-Colonel, and placed in command of this little expedition which consisted of 100 men of his battalion, 200 of another regular regiment, and 200 volunteers. Pursuant to his instructions Polson sailed in January 1780 to Cape Gracias à Dios to pick up a party of Indians and a flotilla of boats which were to join him from the Black River. The flotilla was late, the current strong, the heat excessive; and it was with difficulty that the boats made their way up the San Juan as far as Lake Nicaragua where siege was at once laid to Fort San Juan. But for the arrival of the energetic Captain of H.M.S. *Hinchinbroke*, whose name was Horatio Nelson, it is possible that the lake might never have been reached at all. On April 13 fire was opened against the fort and it capitulated on the 29th; but fever had already played havoc with the attacking force, and the rainy season was at hand. A reinforcement of about 350 men, including 65 of the 60th, commanded by Lieut.-Colonel Kemble of the 1st Battalion, arrived on the scene; but by May 19 the ravages of fever were so great that there were not enough to relieve the guard. Six weeks later 100 men of the reinforcement were dead, and of the remaining 250, ten alone were fit for duty. A second reinforcement arrived, making the total number despatched to the Black River between 1300 and 1400. Of this number less than 400 were alive at the end of August, 58 only being fit for duty. In September the survivors did not amount to 300; in November General Dalling recalled the handful who were left. At least 27 officers had died; four of them belonging to the 60th. It is possible that Polson might have done better, but his faults were venial when compared with the incredible folly of Dalling and Lord George Germaine. The contingent furnished by the 60th had, however, maintained the reputation of the Regiment, and was said to have provided the only good troops in the expedition.

During November General Dalling heard rumours of an attack upon Pensacola. Had not 1000 men been wanted in Nicaragua he might have saved the town; it is, however, fair to say that the mortality in Jamaica was only less than that on the Black River. On September 1 the full strength of his garrison was 3700; within three months 700 were dead. Captain Dalrymple's local garrison regiment was drafted bodily into the 60th. By May 1, 1781, one battalion of the garrison was absolutely extinguished; the remaining seven, including our 1st Battalion, together mustered 2200 men, of whom but 1600 were fit for duty.

It was not until March 1781 that Pensacola was actually attacked. On the 9th of that month de Galvez appeared off the mouth of the harbour with a force of between 5000 and 6000 men brought from Havana. He at once disembarked and invested the town, but undertook no further active operations until joined by 3000 men and a powerful fleet. On April 28 he broke ground, and on May 2 unmasked his batteries. The garrison under Campbell amounted to something over 800 men, including 5 officers and 120 men of the 3rd Battalion of the 60th; part of the remainder being provincial troops. For a week Campbell held his own and made a successful sortie, but the Provincials deserted in large numbers, and one of them pointed out to the Spaniards the principal magazine. De Galvez at once dropped a shell into it, whereupon 76 men of the garrison were killed and 24 others hurt by the explosion. The Spaniards immediately advanced to the assault; their first attack was repulsed, but in a second they established themselves among the débris of the ramparts in such a position that they could pick off anyone attempting to man the guns. Colonel Campbell had then no alternative but to surrender. What the losses were in the detachment of the 60th we do not know; but among the wounded was Lieutenant C. Ward. With the fall of Pensacola the whole of West Florida had been recovered by Spain; and although the remnants of the 3rd and 4th Battalions of the 60th remained for a time in East Florida, the Regiment took no further part in the war. Augustin Prevost went back to England. His brother Marc died of wounds, and the closing scenes of the contest may be rapidly summarised.

In December 1780 Holland joined in the coalition against Great Britain. As far back as May 1778 Lieut.-General Sir Henry Clinton had succeeded Sir William Howe in the command-in-chief; at the instance of the Home Government he had subsequently transferred his principal line of operations to the southern colonies. On May 9, 1780, he captured Charleston; after which, leaving Lieut.-General Lord Cornwallis to reduce South Carolina, he returned to New York. The Americans conducted a harassing guerilla warfare against Cornwallis, which lasted upwards of a year, and the tide of success began to run strongly in favour of the colonists. In April 1781 his Lordship marched into Virginia, and after desultory operations established himself in a fortified position at Yorktown near the mouth of the York River. The security of his position depended upon the British Navy, and at the critical moment the Navy failed him. The consequence was that he was blockaded at sea by the French fleet under Comte de Grasse, and on the land side by an allied force of 16,000 men commanded by Generals Lincoln and Rochambeau. By October 19 hardly more than half of Cornwallis's force of 7000 men remained fit for duty. He then surrendered. Under plea of illness Cornwallis absented himself from the final scene, and it was his second-in-command, Brigadier-General O'Hara, who led out the garrison and directed the troops to lay down their arms. It was O'Hara's fate—a fate it is to be hoped unique in our history—to be twice compelled as a general officer to give up his sword. Twelve years later he commanded the British force defending Toulon, and in an attack by the French was taken prisoner by a colonel named Napoleon Bonaparte.

With the surrender of Yorktown, which would assuredly never have taken place had stout old Prevost been in command, the war so far as America was concerned terminated. In January 1780 Colonel Fuser of the 4th Battalion had died at St. Augustine and was succeeded by Major B. Glazier, who had carried despatches for Braddock in 1755. Early in 1781 the remains of our 3rd and 4th Battalions were sent to New York where the British flag still floated. In June the numbers of the 4th Battalion were 351 of all ranks, of whom 187 only were 'present' at headquarters. Under the terms of the Peace of Versailles concluded in September 1783 the United States gained their independence. The Mississippi formed their western boundary, and among their territory were included certain tracts which had formerly belonged to Canada. But Canada as a whole had been successfully defended by our Colonel-Commandant, Sir Frederick Haldimand, the governor; and now received an accession of strength by immigration from the United States of large numbers of men who had been loyal to Great Britain and had imbibed a bitter hatred for the American Republic. They had good reason. Scoffed at as 'Tories,' they had seen their friends murdered by ardent 'patriots.' They themselves had been subjected to bitter insults and barbarous treatment. They had been driven from their homes, robbed of their property and forced to take refuge in a new dominion. But to us in England the result of their persecution and hardships has been of value beyond the power of words to describe. In the whole Empire there are no more loyal subjects than their descendants; none to whom the link with Great Britain is more precious.

The connection of the 60th with the continent of America, although intermittent and henceforward confined to Canadian territory, lasted for nearly another century.

And so the Royal American Regiment bids farewell to the country in which it was originally raised. To the Regiment the path of duty was plain, and it never shrank therefrom; yet it must have been with much individual heart-burning that the officers and men quitted their kith and kin, and saw them serving under a new flag.

In bidding farewell to its old friends of New England and in congratulating them on independence achieved and future possibilities of incalculable greatness, the Royal Americans no doubt did, as the King's Royal Rifle Corps still does, look back with pride and affection on the country which gave them birth as a regiment, and on the quarter of a century during which their interests were intertwined. Among all the regiments of the British Army, our own corps alone has any historic connection with the United States; and with all due humility we would venture to express the hope that, although long separated, it has never disgraced the American name which it once bore, and that our cousins on the farther side of the Atlantic may still feel not entirely uninterested in its career. Be that as it may, it is undeniable that the first American regiment was that known at the present day by the name of The King's Royal Rifle Corps.

CHAPTER IX

REDUCTION OF THE REGIMENTAL ESTABLISHMENT—THE THIRD AND FOURTH BATTALIONS DISBANDED—RAISED A THIRD TIME—THE FRENCH REVOLUTION—BEGINNING OF THE GREAT WAR—CAPTURE OF TOBAGO—MARTINIQUE — STE. LUCIA — GUADELOUPE — LOSS OF GUADELOUPE

In January 1783 the twelve companies of which every battalion had recently been composed were reduced to eleven, consisting of 80 N.C.O.s and men in the 1st Battalion, and 65 in the others. In June a more sweeping reduction was made ; the 3rd and 4th Battalions were disbanded, the 1st and 2nd reduced to eight companies, including those of grenadiers. The 1st Battalion continued to remain at Jamaica, where its headquarters had been already stationed for sixteen years. The 2nd had been sent two years previously to Barbados, a healthier and in all respects better quarter than Antigua ; but its good fortune was short-lived, for in this very year (1783) it was ordered to send a detachment to Grenada, and in 1785 the remainder of the battalion went to St. Vincent, both islands being covered with forest and saturated with malaria. In 1786 the 1st Battalion, after twenty years' service in Jamaica, was despatched to Halifax, Nova Scotia ; and the 2nd in the following year to Montreal. The regiment was by this time purely English.

Sir Frederick Haldimand and General Augustin Prevost were still Colonels-Commandant of the Regiment, but of those officers who had seen the St. Lawrence before the long exile in the West Indies, George Etherington, now Lieutenant-Colonel of the 1st Battalion, Stephen Kemble, Dietrich Brehm, and J. F. W. Des Barres alone remained.

In 1787 a little difficulty arose with France, who was not only interfering with the affairs of Holland in a manner which gave umbrage to the British Government, but also had a serious difference with it as to the interpretation of an article in the recent treaty of peace stipulating for the French trade in India. The consequence was that each battalion in the army was augmented by one depôt and two service companies. In October the 3rd and 4th Battalions of the 60th were raised for the third time; recruits being obtained partly from England, partly from abroad. With the new battalions entered the Regiment two men, both destined to rise to high command. One became immortal in history, the other notorious; and it fell to the lot of the former to sit as a member of the court-martial which sentenced the latter to be cashiered. The name of the one was John Moore, that of the other, John Whitelocke.[1]

The 4th Battalion was despatched to Barbados; the 3rd had six companies at Antigua, three at Dominica, and two at Montserrat. It now became the practice by sentence of court-martial to send deserters from regiments at home to serve with the 60th in the West Indies, and in March 1789 100 Irish deserters were under orders for shipment. However objectionable this practice to the 60th, the ranks of the 3rd and 4th Battalions were thereby kept up to their establishment, and, says Mr. Fortescue, 'It is probable that the men themselves were much improved. Nothing

[1] The writer has always thought Whitelocke a hardly used man.

is more remarkable than the splendid record of this Regiment in the field at a period when few soldiers entered it untainted by crime.'[1] History is silent as to the doings of the 1st and 2nd Battalions on the St. Lawrence; and in the squalid period of West Indian service to which the 3rd and 4th Battalions were doomed it is difficult to excite any great interest. If it be possible to realise the meaning of the term 'hell on earth,' it assuredly must have been done by those battalions of our Regiment which were quartered in the West Indies. All that the War Office did was to substitute in 1790 loose jackets for coats, and trousers for breeches and gaiters, a trifling improvement which was no doubt accompanied by many protestations that 'the service was going to the dogs.'

Thus sped the few years between the close of the American struggle and the opening of the great war with France. During the interval two distinguished men of our Regiment passed away. The first was Augustin Prevost, who had returned to England a few years before his death in 1786. The second was Sir Frederick Haldimand, who, in 1791, went back to his native town of Yverdun to die.

The little cloud of 1787 being dispersed, the political barometer rose and a period of peace seemed assured, for during the interval that had elapsed since the American war the genius of William Pitt the younger had attracted the minds of the English people to projects of internal progress and commercial development. A high degree of national prosperity had been attained; and Englishmen, forgetting the griefs and disasters of the American war, had been lulled into a state of content and a sense of security, when the French Revolution, sweeping by like a tornado, overwhelmed the current of men's thoughts, uprooted

[1] *History of the British Army*, iii. p. 518.

SIR FREDERICK HALDIMAND

(From the portrait by Lemuel Abbott)

ancient landmarks, and in the exuberance of its fanaticism hurled defiance at United Europe.

The British nation, itself enjoying a large measure of civil and religious liberty, watched the gathering of the storm with some complacency, and regarded the initial efforts of the French reformers with a high degree of interest and sympathy. The principles of freedom professed by both peoples should have led to an alliance calculated to counteract the despotisms of other countries. So far as internal politics were concerned the true course of France in 1789 was to go hand in hand with England in furtherance of the cause of liberty and civilisation. Had such a mutual reliance been practicable it would be difficult to overestimate its effect upon the subsequent history of Europe. Unfortunately the violence of the Jacobins not only put an end to all possibility of such alliance but rendered deadly conflict inevitable. For twenty-two years, except for two brief intervals, the contest was destined to rage; with the ultimate result that both countries narrowly escaped ruin, and that France found herself once more, for the time being, enthralled by the yoke from which at the outset she had freed herself.

At the opening, however, the struggle promised to be but brief. Royalty had indeed bequeathed to the French Republic the splendid legacy of an efficient navy and an excellent army of 180,000 men; but most of the flag officers were either guillotined or driven out of the naval service, while the army, although rapidly quadrupled by the raw levies of revolutionary enthusiasm, seemed at the outset to have but a poor chance of withstanding the forces of United Europe. Indeed, but for the mutual jealousies of the Allied Powers and the incompetence of many of their commanders, one may feel even now a doubt whether the twenty years' war should not have been terminated in as many weeks.

Of England's part in the struggle against Republican France it is unnecessary to speak at length. Her fleet was in a relatively efficient condition, and crews rather than ships were at first wanting and by degrees supplied; but so little had war been expected that when in February 1793 it broke out, a British military force could hardly be said to exist. The only troops at first available for active service were 1700 men of the Foot Guards and a few Artillerymen who, to resist French aggression, were hurried across to Holland in some empty coal barges found at the mouth of the Thames. Three Battalions of the Line followed, but two of them had to be left at the base as unfit for service. During the whole of the year 1793 the purely British force in Flanders barely exceeded 4000 men. Hanoverians and Hessians, bought at so much a head, made up the army to a total of 17,000 men whom the feudal tradition of a Royal Commander-in-Chief placed under the command of the Duke of York, a man who did good work at the Horse Guards in later years, but was quite incompetent as a general in the field. Badly supported indeed he was by his own Government, but on his own account committed every possible blunder; and the conduct of the operations in Holland led the Duke of Wellington, at that time Colonel Wellesley of the 33rd Regiment, to say in after years that he had had the advantage in his earliest campaign of learning exactly how things ought not to be done. Nothing was foreseen, nothing provided for, and the troops perished of cold and hunger. At length, in 1795, the wreck of the army was withdrawn and brought home. But the war continued, and how it was to be brought to a successful issue no one dared say.

But to return to the West Indies. By this time the ideas of liberty with which the French noblesse in general, and the Marquis de Lafayette in particular, toyed during the American War of Independence, had begun to bear wild fruit ten-thousandfold. How the fire of revolution once kindled in France swept the ship of state from stem to stern need not be dwelt upon here. Liberty, equality, brotherhood, excellent in theory, proved something very different in practice. If, as declared by the National Assembly in Paris, all men are free and equal, it becomes obvious that the semi-barbarous negro is as good as the civilised white man. The planters in the French West Indian Islands do not seem to have been greatly attracted by the new doctrine; but their indifference gave place to alarm when they learned that it was seriously proposed in Paris to put into practice the principle enunciated. The negroes on the other hand became wildly excited, and at San Domingo during the winter of 1790–1791, actually rose in rebellion. The revolt was promptly quelled; but supported by their friends in Paris, the blacks rose in insurrection a second time; and taking the planters by surprise massacred several hundreds of them. Anarchy speedily prevailed; cruelty was answered by cruelty, murder by murder, massacre by massacre.[1]

The French planters in the other islands, having, as we have seen, little sympathy with the revolutionary sentiments of their compatriots at home, and fearing a fate similar to that of their friends in San Domingo, now offered to place the whole of their islands in British hands on condition of protection. The British Government warmly accepted the offer, without considering whether

[1] A result of the horror excited in England by this revolt in San Domingo was to prevent Mr. Pitt, the Prime Minister, abolishing the Slave Trade.

the planters were in a position to carry out their promises, or whether their own forces were adequate to give the required guarantee.

But apart from such considerations the West Indian Islands had an important commercial and strategic value. At the period of which we are speaking, nearly one-fourth of the whole import trade of Great Britain was derived therefrom; and in order to secure that trade it was essential to occupy the greatest possible number of the enemy's ports, and so drive out of existence the fleet of privateers which preyed on our commerce.

The West Indian Islands are, roughly speaking, divided into two groups—The Greater Antilles, comprising Cuba and Porto Rico which belonged to Spain, Haiti or San Domingo which belonged to France, and Jamaica, which was the centre of our own trade interests. The Lesser Antilles included a large number of small islands belonging for the most part to England, France, or Denmark. These two groups were also called respectively the Leeward and Windward Islands, from the fact—a most important one before the days of steam—that the Trade Winds blow from N.N.E. to S.S.W., and consequently that although the Windward Islands were in themselves much smaller and of much less value than the Leeward, they had the greater military advantage of being much nearer to the Greater Antilles than the latter were to them. In this group, Antigua, Montserrat, St. Vincent, and Barbados belonged to England; Martinique, Tobago, Guadeloupe, and Ste. Lucia, to France; Trinidad, nearest the South American continent, to Spain.

To deride Pitt's policy of capturing 'sugar' islands is to misunderstand the issues involved. The British navy was now quite strong enough to devote itself to destroying the enemy's privateers; and that being so, strategy demanded the occupation of his ports, and the possession of his islands if only as make-weights in a subsequent treaty of peace. But the point of main importance was the preservation of our trade; and not only was that object achieved, but the proportion of imports increased from 20 per cent. in 1792 to 28 per cent. in 1800. Increase in trade strengthened England's financial stability, the weapon of all others on which she relied for ultimate success. To this end the control of the Caribbean Seas was an object of the first importance.[1] Whether the object was attained, or attempted in the most scientific way is a different matter; and it must be admitted that a terrible lack of economical power, foresight, and regard of human life was displayed almost throughout the contest.

It was decided that Tobago was to be the first scene of operations: this island not being among those to be surrendered. On April 12, 1793, the 4th Battalion of the 60th, 321 of all ranks under Major William Gordon, with two companies of the 9th Regiment, 97 strong, and 50 gunners, sailed from Barbadoes, convoyed by the squadron of Admiral Laforey. At 1 P.M. on the 14th the fleet was at anchor in Courland Bay, and by 3 P.M. the troops landed in Tobago unopposed. General Cuyler, the British Commander, at once advanced against the fort protecting Scarborough the capital, which had been built by the British when in occupation of the island in 1763, and summoned the Commandant to surrender. The summons was treated with contumely, and the General, finding the works stronger than he had expected, halted for the rest of the day.

[1] So clearly did some members at least of the British Government realise the importance of the financial aspect of the war that one of their earliest expedients was the forgery of the enemy's paper money with a view to the depreciation of his credit.

Next morning at 1 A.M. the light companies of the 9th and 60th, under command of Major Gordon who volunteered for the purpose, were detailed to form a storming party supported by the two grenadier companies of the same regiments and the main body of the 60th, now commanded by Captain de Visme; while a party of marines was held in reserve. Firing was forbidden; the place was to be taken at the point of the bayonet. The island is less rugged than most of the others. The advance was undiscovered until, as the column approached the town of Scarborough, a few shots fired by the inhabitants gave the alarm to the garrison at the fort. The guide ran away; and though the column marched straight on up the hill the light companies and part of the grenadiers alone preserved the true direction. As it happened no harm was done: the guns of the fort opened with grape; but while Gordon approached the fort on the weakest side, de Visme came up in the opposite direction against the strongest, and held a part of the garrison in check. Gordon's men got into the ditch, and contrary to the information previously received found the escarp high and perpendicular. Halting his men for a minute's rest, the Major cried, 'God save the King: follow me,' and was answered by three hearty cheers. Up the rampart he clambered followed by a corporal; and the pair pulled up the colours and about a dozen men with the aid of their firelocks. The first four were shot down, but others followed, leaning against the exterior slope of the parapet as they reached it, close to the guns of the fort which were firing over their heads. When the party was complete it charged over the parapet and drove the enemy from his guns without firing a shot. Monsieur Montel, the Governor, then surrendered. The garrison of the place consisted only of 100 regular soldiers and 100

THE CAPTURE OF TOBAGO BY MAJOR-GENERAL CUYLER AND VICE-ADMIRAL SIR JOHN LAFOREY

(From an old print)

sailors, together with a rabble of negroes and mulattoes who probably did more harm than good to their friends; still its capture was a spirited piece of work and a memorable incident in the history of our Regiment; for General Cuyler having been wounded at the outset, Gordon succeeded to the command and gained the victory although himself wounded. The 60th lost 3 men killed and 19 wounded out of a total list of 27 casualties. Captain Maitland of the 60th, D.A.G., was sent to England with the despatch, and a small party of the Regiment was left in Tobago as a garrison.

Martinique was the next point of attack. A force of 1100 men under General Bruce, including three companies of the 3rd and 4th Battalions of the 60th, was detailed for the purpose. The planters had stated that a smaller force would be sufficient to effect the object. The British troops having landed, an attack on the town of St. Pierre was planned for the morning of June 18. Bruce was joined by a column of French loyalists: the force was set in motion, but our French allies fell into confusion and retired from the scene of action. Bruce thereupon re-embarked his men for Barbados. The only casualties were the death of Captain Dunlop and three men of the 60th who had fallen in repulsing a sudden attack made upon the British guns by a party of the enemy during the advance of the 18th. The failure augured badly for the fulfilment of the promises of the French planters.

No other incident occurred during the year excepting the occupation by British troops of St. Nicholas Mole, the best military port in San Domingo, at the instance of the loyalists of that island.

A West Indian campaign on a somewhat larger scale having been decided upon at home, General Charles

Grey,[1] father of the celebrated Prime Minister of 1832, who had done fairly well as a Brigadier in the American war, was selected to command the land forces available for the purpose. He arrived at Barbados on January 6, 1794. The rainy season of the previous year had been more than usually fatal, and battalions were reduced to half or even one-third of their proper strength. The men of the 3rd Battalion of the 60th, with the exception of the grenadier and light companies, were drafted into the 4th, and its Headquarters Staff was despatched to the Channel Islands to recruit the battalion *de novo.* The flank, i.e. the light and grenadier, companies of all the battalions in the Lesser Antilles—comprising probably almost all of their effective men—had already been concentrated at Barbados. They were now formed into three battalions of grenadiers and three of light infantry, the latter being handed over to Brigadier-General Thomas Dundas for training to the standard of the 'perfection of light infantry that was attained during the American war.' The result did great credit to Dundas. The grenadier companies of the two battalions of the 60th together with those of the 15th, 21st, 39th, 43rd, 46th, and 70th formed the 3rd Grenadier Battalion; and the light companies, linked with those of the 21st, 39th, 43rd, 46th, 58th, 64th and 65th, made up the 3rd Light Infantry Battalion.

Barbados being to windward of all the other islands was an excellent base of operations. A descent on Martinique was again decided on, for Fort Royal (now known as Fort de France), its principal harbour, was the most powerful naval station in the Caribbean. Having organised

[1] A grandson of General Grey became at a later date Major in the 60th. He contested and won the borough of High Wycombe in 1830. His opponent was Benjamin Disraeli.

his force of 7000 men in three divisions, Grey set sail on February 3, 1794. His first division under Dundas steered for the Bay of Galion on the east coast of the island; the second under Sir Charles Gordon for Case de Navire; while the third, which he reserved for himself, came to anchor in the Bay of Marin at the south-east corner of the coast. The 60th formed part of the division either of Grey or of Gordon; it is uncertain which. Martinique proved an example of the folly of excessive fortification: in addition to two large forts, Bourbon and Royal, which protected the capital, the whole island was covered with a large number of isolated works. Grey at once decided that the garrison must be frittered away among all these entrenchments, and was probably weak at every point. He disembarked on February 6, at Trois Rivières, a little west of Ste. Luce, and having captured the works by the sea-shore enabled the fleet to sail into Fort Royal Bay, and thus provide supplies for his column. Meanwhile Gordon who had landed on the 8th at Case de Navire, north of Fort Royal, finding the enemy entrenched on the road to the capital, skilfully manœuvred him out of a succession of strong positions by wide turning movements through thick forests. Dundas was equally successful, and by the occupation of Trinité and Gros Morne cut communications between the north and south parts of the island. This effected, he pushed on towards Fort Royal, and on the 14th Grey's combination was successfully effected by the fact of his three divisions coming into touch with each other on different sides of that Fort.

For the moment General Grey contented himself with blockading the enemy's capital with a part of his force, while with the remainder, including the 3rd Light Infantry Battalion, he made another successful combination, resulting in the capture of St. Pierre, the principal trading town of the island.

The General was now able to devote his whole attention to Fort Royal, strongly fortified, but the key to which lay in the command of the neighbouring heights of Sourier, immediately to the north of the town. These heights were already occupied by the French General, Bellegarde; but Bellegarde's mind was set upon taking the offensive, and with that object he descended from his strong position in hopes of attacking the British left and severing its communication with the harbour. Grey countered by reinforcing his left and attacking the heights of Sourier with three battalions, of which the 3rd of Grenadiers was one. The intervening valley was crossed, the roads at the foot of the heights cleared of the enemy's skirmishers, and the steep slopes on the farther side scaled. The heights were won, and Bellegarde who had been repulsed in his attack on the British left received a volley from the British infantry and a salvo from his own guns which had been captured. His men fled in wild confusion, but General Rochambeau, who was in command at Fort Bourbon, denied them admission; and ten days later Bellegarde surrendered with 300 men. On March 6, Grey, who had brought up his siege guns with infinite difficulty, opened a terrific fire upon the garrison of the capital. On the 20th British seamen took Fort Louis by escalade, and the 1st and 3rd Battalions of Grenadiers and 1st and 3rd of Light Infantry having forced their way into Fort Royal Rochambeau had no alternative but to capitulate. On the 23rd Martinique passed into the hands of the British.

During these operations the British casualties did not exceed 350 killed and wounded: the 3rd Battalion of Grenadiers lost 7 men killed and 21 wounded;

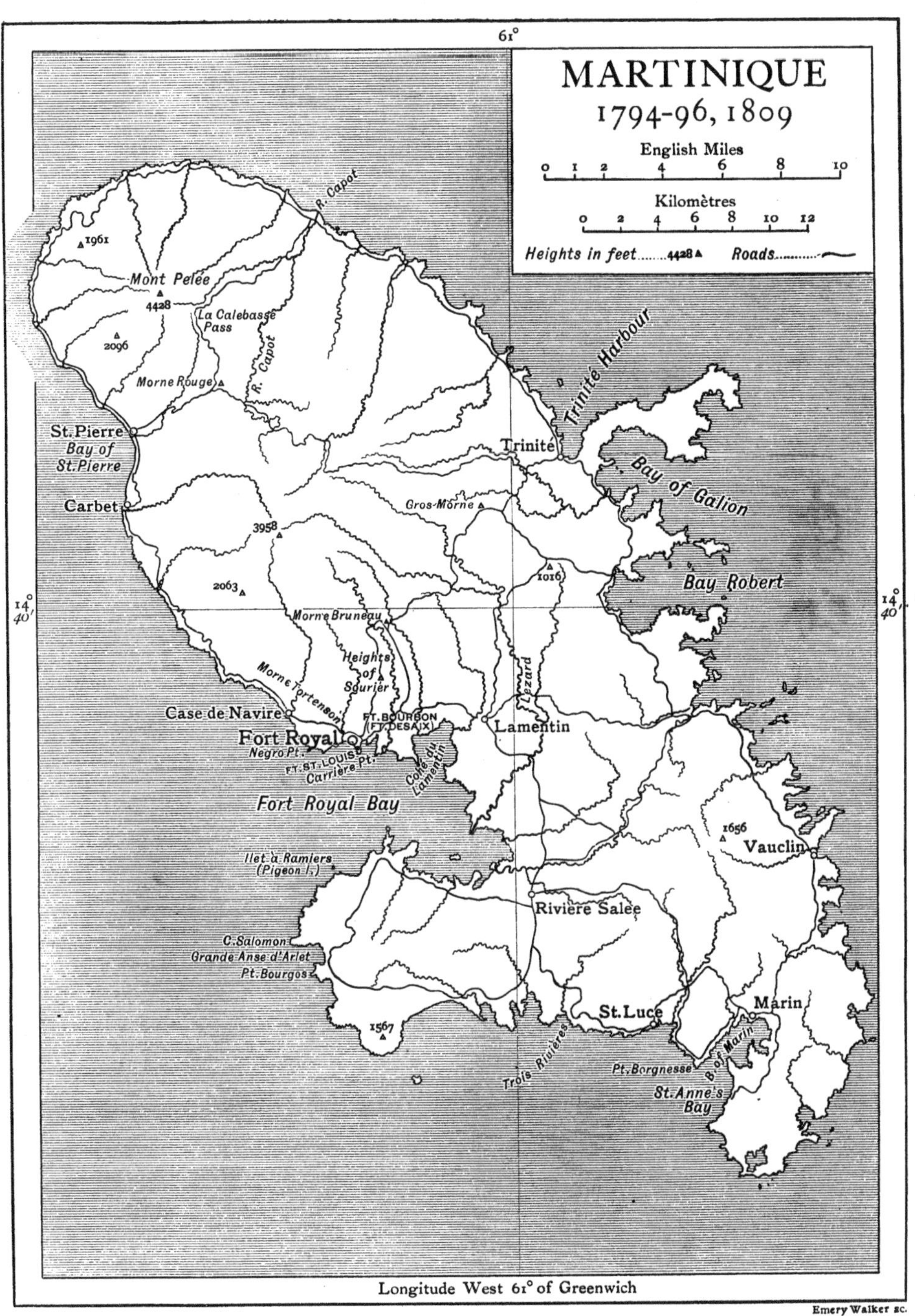
MARTINIQUE
1794-96, 1809
English Miles
0 1 2 4 6 8 10
Kilomètres
0 2 4 6 8 10 12
Heights in feet 4428
Roads
61°
14° 40'
Mont Pelée
4428
1961
2096
La Calebasse Pass
R. Capot
Morne Rouge
St. Pierre
Bay of St. Pierre
Carbet
3958
2063
Trinité Harbour
Trinité
Bay of Galion
Gros Morne
1016
Bay Robert
Morne Bruneau
Heights of Sourier
Morne Tortenson
Lezard
Case de Navire
Fort Royal
Negro Pt.
FT. ST. LOUIS
Carrière Pt.
FT. BOURBON
FT. DESAIX
Lamentin
Cohé du Lamentin
Fort Royal Bay
Ilet à Ramiers (Pigeon I.)
1656
Vauclin
Riviere Salée
C. Salomon
Grande Anse d'Arlet
Pt. Bourgos
1567
St. Luce
Marin
Trois Rivières
Pt. Borgnesse
B. of Marin
St. Anne's Bay
Longitude West 61° of Greenwich
Emery Walker sc.

the 3rd Battalion of Light Infantry 7 killed and 40 wounded—among the latter was Lieutenant Schneider of the 60th.

Leaving a garrison in Martinique, Grey sailed, on March 30, southward for Ste. Lucia. Four columns, landed at different points on the west coast, marched parallel to the shore north and south, took the seaward batteries in reverse, and formed a junction at the fortress of Morne Fortuné which defended Castries, the capital. On the night of April 2 Grey attacked the outworks of the fortress with the bayonet, whereupon the commanding officer and his little garrison of 120 men surrendered.

Having returned to Martinique for supplies, General Grey set sail once more on April 8, this time northward for Guadeloupe, which for practical purposes may be described as two islands divided by a marsh called La Rivière Salée: the part to the east of which is known as Grande Terre, that to the west as Guadeloupe proper. On the 10th the British fleet appeared at the entrance of Point à Pitre harbour, built on the east side of an indentation running back to La Rivière Salée, and during the same night Grey landed a force at Grand Bay on the southern shore of Grande Terre. The landing was opposed by the guns of Fort Fleur d'Epée, which defended the bay, but was, nevertheless, quickly effected. Three columns, each composed of one battalion of Light Infantry and one of Grenadiers, converged upon Fort Fleur d'Epée. A cannon shot from the flagship at 5 A.M. on the 12th gave the signal for the assault. The combination was carried out to perfection; and, after a short but sharp struggle, the fort was carried at the point of the bayonet. The enemy fled to Point à Pitre, leaving behind 60 men killed, 55 wounded, and over 100 prisoners. The British loss included 15 men killed and 60 wounded, of whom 12 killed and 21 wounded belonged to the 3rd Light Infantry Battalion. Captain Robins of the 60th, who was doing duty with the 1st Light Infantry Battalion, was among the wounded.

On the 14th General Grey crossed over to Guadeloupe proper, following the southern road towards Basseterre. On the 17th another column under General Dundas, comprising the 2nd and 3rd Battalions of Light Infantry and the 3rd Battalion of Grenadiers, landed on the west coast seven miles north of Basseterre, and marched on the town. As at Martinique and Ste. Lucia, both columns dislodged the enemy, either by taking the coast batteries in reverse or carrying them by night at the point of the bayonet. On the 20th the two columns formed a junction close to Basseterre, and next day received the surrender of the town.

In these operations Grey's generalship had been of a high order, and at Guadeloupe was indeed brilliant. With a force of probably something less than 2000 men he had vanquished one of at least double his own strength; yet his loss did not exceed eleven men killed, wounded, and missing. The army had worked well with the sister service, which was commanded by an officer whom we last saw at Quebec in 1759, namely the celebrated Admiral Jervis, better known in later days as Lord St. Vincent.

The whole of the Windward Islands being now in our hands, and the object of the campaign attained, the General set to work to administer the conquered island. As summer came on yellow fever made its appearance, and among the first of its victims was General Dundas, the Governor of Guadeloupe, an excellent officer.

On June 5 General Grey, who was inspecting the troops at St. Kitts, was astounded to hear that seven ships, carrying 1500 men, had reached Guadeloupe from France. Not a line had reached him from the British Government to say that this squadron had escaped from Rochefort in April and had been allowed to sail for Guadeloupe unpursued. Thither Grey hurried back, only to find Point à Pitre in the hands of the enemy, though the garrison, consisting of a few companies of the 43rd, had fought most gallantly and been overwhelmed only by superior numbers.

The struggle had hitherto been conducted between French soldiers on the one hand and British soldiers, aided by a few French royalists, on the other; at all events white men had fought against white; but at this juncture Victor Hugues, the French Commissioner, appeared on the scene with a proclamation admitting all black men to equal privileges with white, and summoning all good republicans to rally to him against the English. The result of this proclamation was to reduce the British force to a comparative handful; for the blacks outnumbered the whites by four or five to one, and Grey's original army had been largely reduced by the need of leaving a garrison in the conquered island, and by the ravages of malaria. Hugues was in possession of Grande Terre. He now committed the error of dividing his forces, for he sent a detachment to Guadeloupe with orders to exclude the British squadron from the harbour of Point à Pitre by occupying the headland of Point St. Jean. On the night of the 13th Grey attacked this detachment with the bayonet. Losing only nine men wounded, he captured the enemy's camp, killing 200 men on the spot, and driving the remainder into the harbour where many were drowned.

During the next few days, being reinforced by such small parties of men as could be spared from the neighbouring islands, the General retook the offensive: and, having landed unopposed in Grand Bay, commenced operations on the 19th against Fort Fleur d'Epée. On the 22nd a detachment of Grenadiers surprised a strong post of the enemy at St. Ann's on the south coast, and destroyed the whole of his guns, killing 400 men without any loss whatever. A strong position on the road to Fort Fleur d'Epée was taken in reverse by six companies of Grenadiers and Light Infantry under Lieut.-Colonel Fisher of the 60th after a long night march. General Symes captured Fort Mascotte and repulsed a furious counter-attack on the part of the French. But the struggle was not at an end. More hard fighting followed. The season for hurricanes was at hand, the British troops were exhausted; but Victor Hugues' energy and skill were conspicuous, and it became obvious to General Grey that he must strike home at once or not at all.

By this time casualties had been so many that the Grenadiers and Light Infantry were each organised in two battalions instead of three as hitherto. One Grenadier, and both the Light, Battalions, added to a naval brigade, were detached under Symes to surprise if possible the fortified hill known as Morne Gouvernement which commanded the city; or to penetrate into the town and, by destroying the supplies therein, compel the enemy to evacuate it, as circumstances might dictate. General Grey himself concentrated the rest of his force at Morne Mascotte, with a view to storming Fort Fleur d'Epée on a signal announcing his colleague's success. On the night of July 1 General Symes set out: the darkness was extreme, the road difficult, and the guides unreliable. Finding it impossible to surprise Morne Gouvernement the General directed the battalion upon the town itself. Touch between the units was lost; the sailors went astray, got into the streets, and came under the fire not only of the guns of the Fort which in the first instance they had hoped to surprise, but also of the land batteries and of the ships in the harbour. The seamen were followed by part of the column; officers and men fell on all sides; the whole force got out of hand and fell into disorder. The soldiers who, in accordance with General Grey's instructions, had hitherto relied for success upon the bayonet, now began firing at random, and many who had escaped the fire of the enemy were struck down by that of their comrades. There was no cohesion, no support; it became in the worst sense of the term 'a soldiers' battle.' Batteries were carried at the point of the bayonet; houses were stormed and defending inmates killed; but the whole mass was out of control. Symes was severely wounded. Colonel Fisher probably took the best course in sounding the retreat. Forming a rear-guard of Grenadiers, and aided by reinforcements opportunely furnished by Grey, the Colonel did what he could to collect his scattered parties; and having lost half his numbers, eventually brought the other half to Morne Mascotte. Among the wounded were Captains Foster and Slater and Lieutenant Conway of the 60th.

General Grey said no word of rebuke: he knew how well his force had hitherto fought; but that very fact made him realise all the more keenly the decisive nature of the blow. It was evident to him that his troops were worn out, partly by fatigue, partly by the climate: he made no further attempt to capture Point à Pitre, but retreated to Grande Terre and occupied a strong, fortified position at Berville, hoping to establish himself there until the arrival of reinforcements should enable him to retake the offensive. He then sailed for England, broken down in health, having established a better title to military fame than that of 'the old West Indian marauder,' the epithet which Charles Napier thought fit to confer on him.

The troops left in Guadeloupe remained till nearly the end of the year. Then a miserable remnant of 120 men, who had seen their comrades perish from malarial fever, and had themselves again and again repulsed attacks upon their position at Berville, found no alternative but to surrender: and in Guadeloupe proper, Fort Matilda at Basseterre, which had been defended with equal gallantry, being no longer tenable, was evacuated; the survivors, about 600 men, being safely convoyed to Martinique. Lieut.-General Prescott, in command, gave high praise to the garrison: it had included two companies of our 4th Battalion, which lost one man killed, Captain Walker and 11 men wounded. The fate of the other companies of our Regiment which had been with Grey is unknown. Lieutenant Conway died of his wounds; Lieutenants Cunningham, Schneider, and Cooke of fever;[1] it is even doubtful whether a single man of those companies lived to tell the tale of Guadeloupe. As early as July, thirteen battalions and twenty-eight detached companies of Infantry, together with four or five of Artillery, composing the garrison of the Windward Islands, mustered between them only 5700 men, of whom 1200 were on the sick list. Before the end of the year the force had shrunk to 3200, of whom only 2000 were fit for duty. Of the officers who had sailed from Martinique, 27 had been killed in action; no less than 170 had died of disease. In the Leeward Islands the mortality had been equally great.

[1] It is stated that Lieutenants Montenellier and Best also died of fever. These names do not, however, appear in the *Army List*; but the former may be a misprint for Molesworth and the latter for Bell (Quartermaster), each of whom disappears from the regiment during the year in question.

In Holland the disastrous retreat through all the terrible rigours of winter had been not less calamitous than the yellow fever of the West Indies: and it is little exaggeration to say that by the end of the year 1794 the British Army had ceased to exist. We are never weary of censuring Napoleon for the disasters consequent on his retreat from Moscow; we utterly fail to realise the wholesale manner in which British troops have again and again been sacrificed by purblind ministers and incompetent commanders.

CHAPTER XI

CARIB REVOLT AT ST. VINCENT—SIR R. ABERCROMBY ASSUMES COMMAND—THE FIRST AND SECOND BATTALIONS—ABERCROMBY'S OPERATIONS—DEATH OF LORD AMHERST—FIFTH AND SIXTH BATTALIONS RAISED—RIFLE COMPANY OF THE SIXTH BATTALION AT EGMONT-OP-ZEE—PEACE OF AMIENS—DESTRUCTION OF THE FRENCH ARMY AT SAN DOMINGO

VICTOR HUGUES, to whose vigour and audacity must be credited the expulsion of the British from Guadeloupe, seems, as remarked by Captain Mahan in 'The Influence of Sea Power,' vol. i. p. 116, 'to have embodied in himself the best and the worst features of the men in the Terror. . . . In his report to the Convention he boasted of having put to death 1200 royalists in Guadeloupe.' He was undoubtedly a man of determination and ability: his idea was to make Guadeloupe a centre of revolt and a base whence men, arms and supplies of all kinds could be forwarded to the insurgents of the neighbouring islands. The timely energy of the French Government supplied him in January 1795 with a much-needed reinforcement of naval and military force, which was but partly counterbalanced by the arrival of three battalions with which the British Government had somewhat earlier provided Grey's successor, General Vaughan. The fame of Hugues rapidly spread. At St. Vincent, early in March,

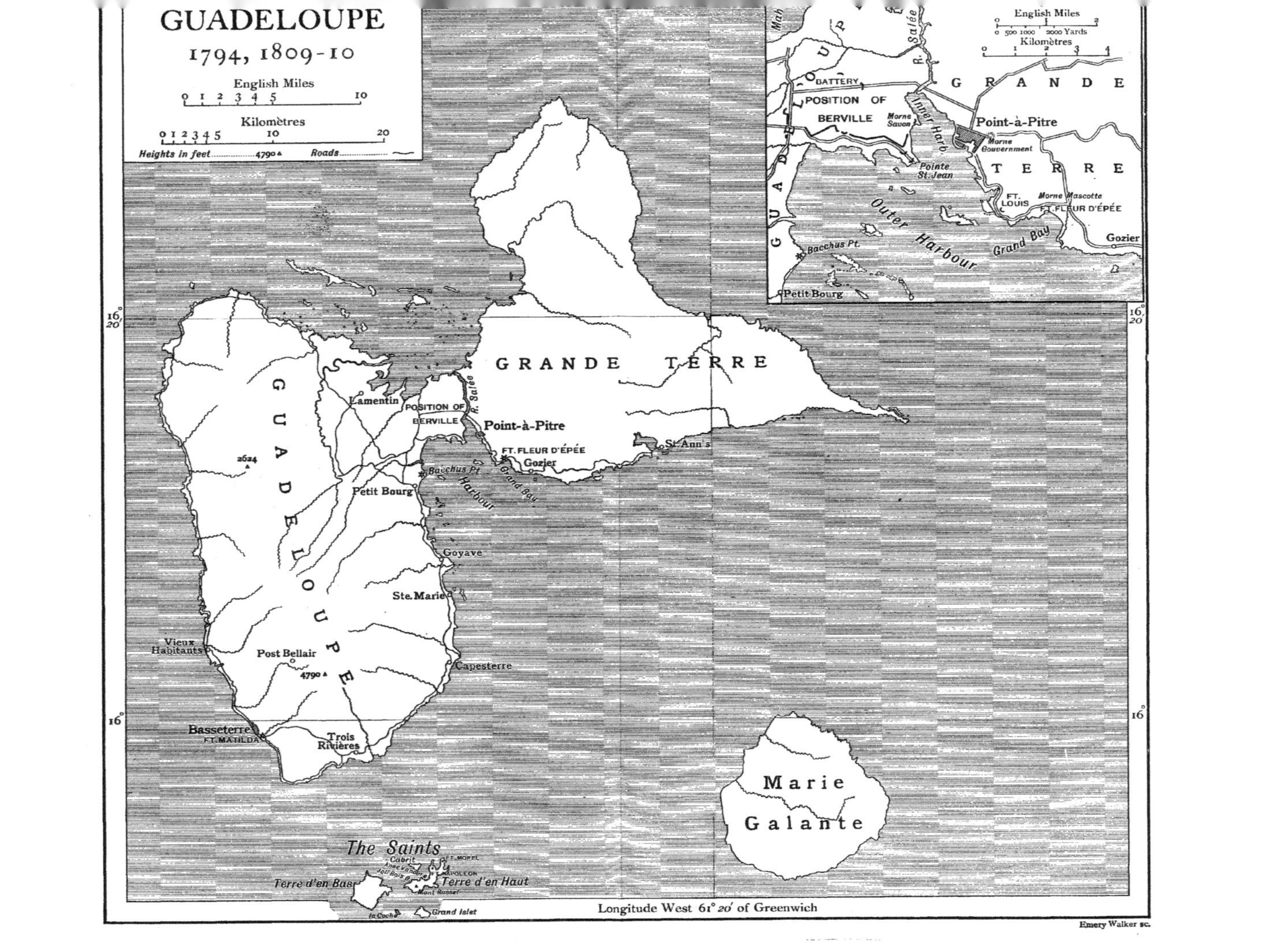
GUADELOUPE
1794, 1809-10
English Miles
Kilomètres
Heights in feet 4790
Roads
GRANDE TERRE
GUADELOUPE
Lamentin
POSITION OF BERVILLE
R. Salée
Point-à-Pitre
FT. FLEUR D'ÉPÉE
Gozier
St. Ann's
Bacchus Pt.
Harbour
Grand Bay
Petit Bourg
2624
Goyave
Ste. Marie
Vieux Habitants
Post Bellair
4790
Capesterre
Basseterre
FT. MATILDA
Trois Rivières
Marie Galante
The Saints
Cabrit
Terre d'en Bas
Terre d'en Haut
Grand Islet
Longitude West 61° 20′ of Greenwich
Emery Walker sc.
16° 20′
16°
English Miles
Yards
Kilomètres
BATTERY
POSITION OF BERVILLE
Morne Savon
Inner Harb.
Point-à-Pitre
Morne Gouvernement
GRANDE TERRE
Pointe St. Jean
FT. LOUIS
Morne Mascotte
FT. FLEUR D'ÉPÉE
Outer Harbour
Grand Bay
Gozier
Bacchus Pt.
Petit Bourg
GUADELOUPE

the Caribs rose in insurrection. The handful of men, partly invalids of our 4th Battalion, partly of the 46th, who composed the garrison, were driven into the capital. At Martinique a simultaneous rising occurred: white men were massacred by black; others, including the Governor, were captured and held as hostages. At Grenada, as at St. Vincent, the island, with the exception of the capital, was lost to the British. In June the English garrison evacuated Ste. Lucia. It looked as if the British flag was to be expelled from all the Windward Islands. Every available man was surely wanted; yet the British Government had taken this opportunity to send to Demerara the 3rd Battalion of the 60th, whose ranks had recently been filled up, though by foreign recruits of the worst class. It had been hoped that the Dutch would yield Demerara without resistance: the hope was falsified, and the 3rd Battalion was for months kept cruising about on board ship at a time when every available man was required to keep the British flag flying in the Eastern Caribbean. At length in June, the battalion, having returned to Barbados, was despatched thence to Kingston, the capital of the island of St. Vincent, where desperate fighting had taken place. On its arrival it formed part of a force under Colonel Leighton of the 46th, about 800 strong, collected to clear the windward side of the Island. On the 12th Leighton drove the enemy from Vigie, a strong position, after a hard fight in which Captain Piquet of the 60th was killed and Lieutenant Tonson wounded. The column then advanced to Mount Young, which it fortified, although such was the enervating character of the climate that it took three days to march a distance of less than twenty miles. But short as was the march, no less than eight men died of exhaustion on the third day. Leighton fortified Mount Young and the adjoining peak, pushing out patrols on the north side, with a view to depriving the Caribs of their fertile valleys, and of forcing them to starve in the mountains.

Up to the present, matters had gone fairly well, but the loss of Ste. Lucia—the island immediately to the north—began to exercise a baneful influence upon our operations. Utilising that island as a base, the French were able to fan the flame of revolt in St. Vincent by sending thither men and stores as required. Early in July a party of French and negroes took post opposite Château Belair on the west coast. Seton, the Governor, with more zeal than discretion, directed Lieut.-Colonel George Prevost of the 60th to dislodge them with 30 men of his battalion and 70 blacks. The attack, delivered against greatly superior numbers, failed: the leeward side of the island was now at the mercy of the enemy; and Seton, unable to get reinforcements from Martinique, felt obliged to weaken Leighton by a demand for a part of his force. On August 5, 200 men, partly of the 60th, partly of the 46th, supported by 100 men of the native militia, made a second attempt upon the enemy's position at Château Belair. From midnight until 8 A.M. the British force had marched through deep and airless woods and ravines; nevertheless they captured the enemy's works right gallantly but with the loss, proportionately great, of 60 officers and men killed and wounded. The greater part of the fighting fell to the 46th, and to that regiment the greater credit is due. The casualties of the 60th were 2 officers and 13 men; the insurgents retired with the loss of two guns and many men.

In hopes of stopping reinforcements from reaching Ste. Lucia, Governor Seton established a negro post at Owia, which commanded the little harbour at which the reinforcements disembarked; but the distance between Owia and Kingston is five-and-twenty miles, and communication between them had to be maintained by detachments, small in themselves, but which frittered away the force at Seton's disposal. Yellow fever had happily been so far entirely absent, but exposure to the climate of men destitute of shoes and ragged in clothing could not fail to make ravages in our ranks and those of the 46th ; the 3rd Battalion and two or three companies of the 4th had but 400 men who by any stretch of imagination could be thought fit for duty. In the first week of September the French surprised Owia and once more poured reinforcements into the island. Then, having established themselves in force, they advanced, and on the 24th occupied a ridge commanding the road from Vigie to Kingston. Colonel Myers, now in command of the troops at St. Vincent, concentrated such force as he was able. Vigie still held out ; but famine threatened ; and Myers having collected supplies, ordered Lieut.-Colonel Ritchie, who had recently joined the 60th, to take 200 men of his battalion and the same number of negroes to escort a convoy to revictual Vigie. In pursuance of his order Ritchie advanced, and was nearing his destination when he found the road in the occupation of the enemy. A party of the 60th was ordered to the front to clear the way, but the enemy was in force and held his ground so strongly that the detachment was driven in upon its supports. Ritchie's men were exhausted and disorganised by hunger, and the whole column fell back in disorder. The convoy fell into the hands of the enemy ; but the Colonel, although wounded, collected a body strong enough to repel any attempt at pursuit, and the officer in command at Vigie was sufficiently adroit to bring his men away during the course of the night. Misfortunes such as these will infallibly occur when a regiment, as in the present instance, is filled up with men of the lowest character, unused to discipline, devoid even of the tie of national feeling, or the opportunity of becoming imbued with the training and spirit of the corps in which they were enlisted. The wonder, as the Hon. John Fortescue observes, is 'not that these 3rd and 4th Battalions of the 60th on occasion did badly, but that on the whole they did so well.' We have seen that, by the end of the previous year, the British Army had to all intents and purposes ceased to exist: it was impossible to create a new one in six months.

'Service in the West Indies,' observes Mr. Fortescue, 'was loathed to an extent which is hardly credible in these days; and this was only natural, for there was no glory to be won, no profit to be gained; the work was most difficult and dangerous, the hardships very great, the climate deadly and unwholesome, the arrangements for clothing chaotic, the organisation of the hospitals abominable, and lastly the Government's ideas of carrying on war were hopelessly imbecile.'

Early in October Major-General Irving, with a reinforcement of three battalions from England, arrived to take command. The enemy was dislodged from his position at Vigie, but within three months the post was recaptured, and the troops became thoroughly demoralised. On January 20, 1796, Colonel Prevost projected an attack upon an isolated picquet of the enemy: although he often did well in the field, his subsequent career does not give the impression that Prevost (a son of Augustin) was a great leader of men; anyhow on this occasion he was unable to inspire a detachment of his battalion with the necessary confidence, and the attempt was abandoned. The French attacked in their turn, and after a fight which lasted all day were repulsed with heavy loss. The position was, however, very serious, for it became evident that the troops could not be relied upon in any sustained effort.

General Vaughan, Commander-in-Chief of the Windward Islands, complained to the British Government that he had no troops in the real sense of the word, merely sick men and boys; but as he realised that the Government had literally nothing else to send he begged to be allowed to form negro regiments. 'It is only filling the hospitals,' he wrote home, 'and deceiving yourselves to send out raw or newly-trained levies.' His idea was no doubt correct, and has since been carried out with considerable success; but at that time the British Ministry was not prepared for such a step, and to their refusal much of subsequent disaster may be traced.

Although Pitt and Dundas, the Secretary of State, were unwilling to carry out Vaughan's suggestion, they realised how serious was the situation, and made what it is only fair to term a great effort. They entrusted the Command-in-Chief to Lieut.-General Sir Ralph Abercromby, the best officer among the senior ranks of the British Army; and associated with him as Brigadier-Generals the very best men they could find, namely John Moore, John Hope, and Knox, of whom the first two had during their career a temporary association with the 60th. Abercromby was given no less than twenty-two battalions, mustering nearly 18,000 men, the greater part of which, after meeting the worst of weather in the Atlantic, was concentrated at Barbados by the end of April. In numbers the Army was respectable; in quality it could hardly have been worse, but it contained a good battalion of German riflemen, designated Löwenstein's Chasseurs, which, two years afterwards, was embodied in the 60th. Chiefly through the good conduct of this corps Ste. Lucia was captured in May, and the insurrection at St. Vincent practically terminated by the capture of the position at Vigie on June 9.

What remained of the 4th Battalion was sent in the autumn to the Channel Islands where it was quickly filled up to at all events a nominal establishment of 1100 men, and in the following year, 1797, returned to the Windward Islands. The 3rd Battalion had in the meanwhile been quartered at St. Vincent and Tobago.

History at this period has little to relate about the 1st and 2nd Battalions of the 60th which, during the last ten years, had been pursuing the even tenor of their way on the St. Lawrence, with their headquarters for the most part at Montreal and Quebec.

In 1794 a battalion of the Regiment, apparently the 1st, was officially armed with the rifle. We have seen that, as early as 1758, a few rifles had been experimentally issued to Colonel Bouquet's battalion, and there is evidence to show that officers of the Royal Americans and some at least of the men habitually carried rifles when engaged in frontier warfare. The weapon was common enough in America, but this issue of 1794 was the first made on a large scale to a regiment of the British Army by order of the War Office. At or about the same time a Rifle Company was formed in each of our other battalions.

It is probable that these two battalions, left on mere garrison duty and not employed in the West Indian operations, were allowed to fall to a very low establishment; at all events we know that in June 1797 the 1st Battalion was drafted into the 2nd at Quebec, and its Headquarter Staff despatched to Guernsey. At this island was collected a considerable number of ex-soldiers attached to the monarchies of France and Holland, and unwilling to serve the Republican Governments. With these men (many no doubt extremely valuable soldiers) the ranks of the 1st Battalion were speedily recruited. Before the end of November the battalion re-embarked at Portsmouth; and for some reason unknown, but which probably had reference to contrary winds, disembarked again. The fear of French invasion and the Irish Rebellion of the following year combined to detain it at home, and it was not until 1799 that the battalion sailed for Barbados.

At that island in the first weeks of 1797 Sir Ralph Abercromby had concentrated 4000 men, which included four companies of the 3rd Battalion of the 60th, detachments from Löwenstein's Chasseurs, Löwenstein's Fusiliers, and Hompesch's (another foreign) Corps.

In October 1796 Spain, elated by her recent alliance with France, had declared war upon England.

'The alliance of Spain and Holland with France,' observes Mahan, 'increased the difficulties of Great Britain by throwing open their colonial ports to French privateers. The extensive sea-coasts of Cuba and Hayti (San Domingo) became alive with them.'[1]

The orders which Abercromby had received were to capture Porto Rico or Trinidad as he should judge best, but preferably the former, and to transport thither the French royalist families from St. Domingo whom the British Government had engaged to protect; for the prospect of subduing the negroes in an island of that size appeared to be very distant.

Abercromby sailed from Barbados on February 15, 1797, for Trinidad; two days later he landed close to the Port of Spain, and on the 18th the island was surrendered by the Spanish Governor. Leaving a garrison at Trinidad, Abercromby returned for reinforcements to Martinique, whence on April 8 he embarked with a force, still of 4000 men, for Porto Rico, and on the 17th anchored off the town of St. Juan.

Having disembarked his troops the General made an attempt to seize the city. But its defences were too

[1] *Influence of Sea Power upon the French Revolution*, i. p. 120.

strong, and after losing 150 men Sir Ralph re-embarked for Martinique. The 3rd Battalion returned to Tobago.

The name of Amherst at this period seems to connect us with a bygone age; yet all this while Sir Jeffery—now Field-Marshal Lord—Amherst was still alive. He lived till August 23, 1797, having been for thirty-nine years Colonel-in-Chief of our Regiment. It so happened that he died just too soon to see the new departure which by degrees changed the colour of our uniform and gave us officially the name of a Rifle Corps. In December an Act was passed to enable a fifth battalion to be added to the 60th. On January 12, 1798, an order was issued for the new battalion to be raised; it was armed with rifles,[1] dressed in green, and placed under command of Lieut.-Colonel Baron Francis de Rottenburg, an experienced Austrian officer of light infantry, who at an earlier date had been in the French service. The story of this battalion will be told in another volume. Amherst was succeeded in the Colonelcy-in-Chief, as well as in command of the army, by H.R.H. Frederick, Duke of York, whose appointment must be taken as a distinct compliment to the Regiment.

The 1st Battalion was inspected at Barbados in March by General Trigge, who reported it to be a very fine battalion and in excellent order. Here it was joined first by two companies and afterwards by four more of the 2nd Battalion from Canada. The headquarters of that

[1] The rifle was, as we have seen, known to the regiment as far back as 1758. In 1776 an improved weapon was exhibited to Lord Amherst by a Major Ferguson at Woolwich. Ferguson fired first four, then six shots a minute. He hit the bull's-eye of a target 200 yards distant, lying on his back. The exhibition we are told excited great admiration.

battalion remained at Quebec, or possibly Montreal; but two other companies were sent to Trinidad, whose Governor was Colonel Thomas Picton, at a later date Wellington's well-known General of Division.

An Act of Parliament, dated July 12, 1799,[1] provided for the augmentation of the Regiment by a sixth battalion with an establishment of 1159 of all ranks. This Act had been to some extent anticipated, for as early as February in the same year a company had been formed, dressed in green and armed with rifles. During the month of September this company embarked on the *Camilla* frigate for Holland; and about the 25th joined the army commanded by the Duke of York, which was already engaged with the French in the campaign usually known as 'the Expedition to the Helder.' A Russian force acted in combination with ourselves. On October 2 an attack was made on the enemy's position at Egmont-op-Zee: the company of our 6th Battalion formed part of a brigade under Colonel Macdonald attached to the division of Sir Ralph Abercromby. Macdonald neglected the orders which he had received to protect Abercromby's left flank during the advance; followed up some parties of the enemy; got entangled among the sandhills; fought an independent action on his own account, and eventually joined Abercromby at dusk with the jaded remnant of his four battalions. The loss of the rifle company amounted to 6 killed, 7 wounded, and 4 missing. The result of the whole action was indecisive, and the army shortly afterwards evacuated Holland. The rifle com-

[1] During this year a book called the *Regimental Companion* was published by Charles James, an officer of the Regiment. The book is a sort of compendium of military life and duties. The writer was far in advance of his age; and the reforms which he advocates—abolition of purchase and of flogging, &c.—were not carried out for seventy years.

pany thereupon joined the headquarters of its battalion at Parkhurst in the Isle of Wight. The following is a copy of its 'Monthly Return' dated June 30, 1800:

	Lieut.-Colonel	Major	Captains	Capt.-Lieut.	Lieutenants	Ensigns	Paymaster	Adjutant	Quartermaster	Surgeon	Assist. Surgeon	Sergeants	Drummers	Rank & File
Present, fit for duty	1	1	5	—	9	5	1	1	1	—	1	44	17	700
Sick in { Quarters .	—	—	—	—	—	—	—	—	—	—	—	—	—	25
Sick in { Hospital .	—	—	—	—	—	—	—	—	—	—	—	—	—	13
Absent with Leave .	—	—	—	—	1	—	—	—	—	—	—	9	5	262
Not joined since appointment, wanted to Complete .	1	—	2	1	1	7	7	—	—	—	—	—	—	—
Joined . . .	—	—	—	—	—	—	—	—	—	—	—	—	—	53
Dead . . .	—	—	—	—	—	—	—	—	—	—	—	—	—	2
Discharged and not recommended .	—	—	—	—	—	—	—	—	—	—	—	—	—	4
Deserted . .	—	—	—	—	—	—	—	—	—	—	—	—	—	5

In October the battalion was inspected by the Duke of York, and in the following month embarked for Kingston, Jamaica.

A year or two before this time the British Government had decided to carry out General Vaughan's suggestion, by raising twelve battalions of negroes under the name of West Indian Regiments. The measure was undoubtedly wise and beneficial; but that it did not seem so to the white planters who had seen the massacre of Europeans by blacks at San Domingo is no matter for wonder. Wild panic and impotent rage ensued: the Assembly of Jamaica with some reason decided to pay and feed British regiments for their own protection. To this end the 1st and 4th Battalions of the 60th were to be augmented to an establishment of 800 and 1200 respectively, and to be quartered in the island. The increase was never carried out in its entirety: in October 1799 the 1st

Battalion mustered but 640 men, and the 4th, 690. In March of the following year the alarming situation at San Domingo induced the British Government to send a reinforcement to Jamaica of two battalions, one as already noted being the 6th of the 60th. Their arrival however was delayed; they did not reach the island till March 1801, by which time the customary ravages of the climate had reduced our 1st and 4th Battalions to a joint strength of hardly more than 900 men. After a time the 6th Battalion also lost many men from the same cause, but at the outset it was more healthy; and the period was now happily approaching when at the dictates of elementary common sense the barracks were removed to the cool and healthy mountain summits of the island. By this time the first stage of the great war was almost at an end. Before the close of 1799 the voice of the French nation had called to supreme power, under the designation of First Consul, General Napoleon Bonaparte, who had just returned home after his victorious career in Egypt and in whose absence matters had gone badly for the French Republic. The first act of his power was an endeavour to bring about peace with England. But the idea of re-establishing the Bourbon régime in France was still clung to by her enemies, and it is curious that although General Bonaparte had done no harm to England—had indeed hitherto taken no part in politics—his accession to office elicited from the ruling classes in Great Britain an outburst of personal hatred which had never been shown even to the most sanguinary participators of the Reign of Terror. His overtures were in consequence received with contumely, almost with insult, and for the moment bore no fruit. His victory at Marengo in 1800, and that of Moreau at Hohenlinden, tended, however to bring his foes to a more reasonable frame of mind. Peace on the Continent was secured by the Treaty of Lunéville of February 8, 1801. In March British *amour propre* was flattered by the successful campaign of Abercromby and Hutchinson in Egypt; everyone was heartily tired of war, and in October a preliminary treaty was signed, which, in March 1802, resulted in the Peace of Amiens, by the terms of which our conquests in the Windward Islands were restored to France.[1]

But in the West Indies there was no rest. General Bonaparte, feeling it essential to re-establish his power in San Domingo, an island absolutely essential to his West Indian commerce, sent thither a force of 32,000 men. His General, Le Clerc, whose first regiments landed in February 1802, at the outset achieved some success against the blacks; then the climate did its deadly work. In October a French aide-de-camp arrived at Jamaica with the request to the Governor, General Nugent, for a loan of £60,000 to carry on operations. Nugent extorted from him the admission that more than five-sixths of the French army was already dead. The alarm with which the French enterprise had at first been greeted had by this time disappeared; to all appearance but a few months were required for the utter annihilation of Le Clerc's forces. Winter passed. On April 28, 1803, our 2nd Battalion was at St. Vincent, the 3rd at Grenada; but in June 1803 news arrived that war had once more broken out between France and England. The British Admiral, Duckworth, lost no time in blockading the French fleet and the maritime posts still occupied by the French in San Domingo. Their supplies cut off by the British squadron, their forces almost overpowered on land by the negroes, the condition

[1] The evacuation of the restored islands took place as follows: On September 12, Admiral Villaret Joyeux took over Martinique. Tobago was given up early in October, Ste. Lucia having been restored a few days earlier. Demerara was given back in December.

of the unhappy French was desperate. Throughout the whole of the war nothing is more horrible than the way the British treated their French prisoners; yet, even so, to fall into the hands of the British was perhaps a shade better than to be exposed as prisoners to the barbarities of the blacks; anyhow the French quickly made their choice, and gave themselves up in large numbers to Duckworth and his officers. Bonaparte's visions of a West Indian trading centre were at an end.

Jamaica now became crowded with prisoners of all nationalities; for the First Consul, knowing the French repugnance to over-sea service, had incorporated among the regiments sent to San Domingo a large number of Poles, Swiss, Germans, &c. The loyalty of the Polish contingent seems to have been rather doubtful; many, it is said, actually deserted to the negroes and fought their own comrades. Others were brought to Jamaica as prisoners, and 400 of them being perfectly ready to enter the British service, were drafted into the 60th. The history of their subsequent service is unknown, but at all events the regiment can claim to have received a draft from the great Napoleon himself!

CHAPTER XII

RUPTURE OF THE PEACE OF AMIENS—EXPEDITION TO GUIANA—THE SECOND BATTALION IN SPAIN—CAPTURE OF MARTINIQUE—GUADELOUPE—THE UNITED STATES DECLARE WAR—THE REGIMENT AS MARINES

THE title given at the outset to the new ruler of France was that of 'First Consul,' but to all intents and purposes General Bonaparte was an absolute monarch. It must, indeed, have become evident long ago to all Frenchmen of observation and reflection that the so-called constitutional government of the Revolution had utterly failed and that all hope for the future lay in the beneficent despotism of a single ruler. The work of the Revolution had been entirely destructive; the task set himself by Bonaparte was purely constructive. His object was to heal his country's wounds, and by establishing France upon the basis of commercial prosperity to give her a new lease of life. It is not by reading the record of his military exploits, splendid as they are, but by studying his career as a civil statesman that one grasps the aims and ideals of his life. The measure of his success is shown in France by his monumental achievements such as the Code Napoléon, the Concordat, the Cadastre; by the colleges which he founded; by the roads and canals which he constructed, by the deserts which he afforested, and in so doing forced to yield an income to the State. On the Continent he also did great things. How far he would have gone but for thc inveterate enmity of Great Britain can only be matter for conjecture, for the naval power of his foe stopped his projects for oversea trade and colonial expansion.

The Peace of Amiens proved but an armistice. Each side with reason felt it had causes for grievance against the other. On February 20, 1803, the First Consul made an address to the Legislature of an eminently pacific and moderate character. Unfortunately he used therein an infelicitous expression which, wrested from the context, gave dire offence in England whose Government refused to carry out its promise to evacuate Malta. A quarrel of words ensued. Bonaparte was bent on the continuance of peace, for his mind was full of schemes for the internal reform of his country and for her civil and commercial development. He neither wished nor expected a rupture but expected to gain his point by the adoption of a firm attitude. To his surprise the British Ministers, on May 1, closed the discussion by a declaration of war.

When one remembers that the ruler had already proved himself by far the most able military commander in Europe since the days of Marlborough, and that the resources of the country were at the disposal of one not less great as an administrator than as a general, it becomes obvious that, for the impending encounter, Britain needed all her statesmanship, all her resources, and her entire concentrated strength.

A titanic struggle ensued, involving at different periods every country in Europe. But France and England were the principals, making use of the Continental Powers as puppets. Neither country would permit neutrality; and the sovereigns of Europs, *nolentes volentes*, were compelled to join in the struggle, sometimes on one side sometimes on the other, as France or England alternately predominated.

In order to appreciate the part taken by our Regiment it may be well to give a slight sketch of the salient features of the contest.

The object both of France and England was commercial supremacy.

The idea of Napoleon Bonaparte was to bring about the downfall of Britain either (1) by the invasion of England, (2) by breaking her power in the East, or (3) by prohibiting the entry of her commerce into the Continent of Europe.

The first and second of these projects were ruined in October 1805 by the annihilation of his battleships at Trafalgar. In the third, Napoleon was within measurable distance of success, yet it ultimately proved a leading cause of his own ruin.

The plan of Great Britain was to become the sole mistress of the seas and to crush the military power of France by subsidising against her the continental hosts of Europe.

She almost achieved the first aim at Trafalgar, yet could never entirely rid the seas of the enemy's privateers which continued to prey upon her commerce, and were in 1812 reinforced by those of the United States.

In the second she may be said to have had ultimate success, but not without experiencing many bitter failures and disappointments.

As soon as war broke out Mr. Pitt (son of the William Pitt better known as Lord Chatham, whom we met at the outset of our regimental career) turned his attention to the formation of a coalition which should coerce France by weight of military power. To this coalition Russia,

Sweden, and Austria gave their adhesion; but the decisive defeat of the Austrian and Russian armies at Austerlitz in December 1805 forced Austria to make peace and Russia to retreat. During the following year Prussia declared war on France, but in October was broken to pieces at the battle of Jena.

In June 1807 the terms of the Treaty of Tilsit not only ended the war between France and Russia, but the two Empires—for in December 1804 General Bonaparte had been proclaimed Emperor of the French, under the title of Napoleon I—entered into a close alliance. Great Britain found herself left not only to continue the contest with France single-handed, but, nominally at all events, with the whole of Europe against her also.

But during Napoleon's Prussian campaign a manifesto had been issued by Spain with the undoubted intention of being followed up by a declaration of war against France. The possibility of being attacked by his southern neighbour was fraught with danger to Napoleon, the more so that France and Spain were nominally in alliance. The affront was one which the Emperor could not fail to resent; and although news of the battle of Jena immediately caused the manifesto to be explained away, yet he took steps, as soon as the Treaty of Tilsit gave him leisure, to prevent a recurrence of the danger. His primary object was, however, to get possession of the fleet of Portugal, at this period England's sole ally. To this intent a French force, acting in alliance with Spain, occupied Lisbon; but too late, for the Portuguese men-of-war were already clearing the harbour. Foiled in his immediate object, Napoleon was determined to retain an effective footing in Spain, and by equivocal conduct gained possession of her principal fortresses in the north. The internal circumstances of Spain aided his plans. The King and his son, realising for the moment their own incapacity, in turn relinquished in favour of the French Emperor their claims to the throne, to which Napoleon, supported by a large and influential body of Spaniards, proceeded to appoint his brother Joseph. But the mass of the people utterly refused to recognise the arrangement, and an insurrection throughout the length and breadth of the land rapidly followed.

Great Britain, at this period, was nominally at war with Spain; but her aid was invoked, and her despatch of troops initiated what is commonly known as the 'Peninsular War.' In 1809 Austria was again induced to take up arms against France, but was again compelled to conclude a disastrous peace.

During the following years the power of Napoleon was at its height. He hoped to ruin England by placing an embargo upon her commerce throughout the whole of Europe. But the difficulty of excluding British trade from the Continent increased year by year. Russia could at length stand it no longer. In the summer of 1812 war broke out between her and France. Napoleon crossed the Niemen, and during the autumn occupied Moscow with an army sadly reduced by the effects of battle and disease. In his subsequent retirement, cold and hunger were added to the weapons of his enemy. A comparatively small proportion of his army recrossed the Russian frontier.

In the following year, first Prussia, then Austria and the South German States joined the forces of Russia, and the campaign of 1813 ended in the occupation of the south-western departments of France by an Anglo-Portuguese army under Wellington, and of the provinces on the left bank of the Rhine by the allied Powers of the north.

In 1814 the superb skill of Napoleon failed to prevent

the capture of Paris, and on April 6 he relinquished his throne and retired in exile to Elba. Recalled in the following year by the voice of the French nation, Napoleon once more ascended the Imperial throne, from which however, he was again shortly driven by the combined Powers of Europe, cemented by the victory of Waterloo. And so ended the great war.

The year 1803 finds the 1st, 4th, and 6th of our battalions at Jamaica; the 5th at Surinam. During the last few years great changes had taken place in the *personnel* of the officers. But a single survivor of the regiment as originally raised existed in the person of Joseph Des Barres who, at the comparatively mature age of eighty-one, was still borne on the list of his battalion as a captain, although with the brevet rank of colonel in the army. Among the lieutenant-colonels we find the name of George, son of Augustin Prevost; and among the majors, those of Edward Codd, destined to command the 2nd Battalion for an almost unprecedented number of years, and of John Keane, who nearly forty years later battered down the walls of Ghuznee. In the list of officers we also find the name of Charles de Saluberry, the hero of Chateauguay in 1813.

In the West Indies the rupture between England and France had been expected for some months before it actually took place. At the end of April 1803 Lieut.-General Grinfield, General Officer Commanding the Windward and Leeward Islands, received orders to be prepared for all contingencies. The troops under his command comprised between ten and eleven thousand men of all ranks, about two-thirds of whom were Europeans. Nearly three thousand men were at Barbados: the remainder had been about equally distributed between the islands of Trinidad, Grenada, St. Vincent, Barbados, Dominica, and St. Kitts—the 2nd Battalion of the 60th and the 1st West India Regiment forming the garrison of St. Vincent; the 3rd Battalion and the 3rd West India Regiment, that of Grenada.

On June 20 Grinfield received definite information that war had broken out with France. The French islands of Ste. Lucia and Tobago were at once captured, but without the aid of the 60th. A month later came news of war with Holland, a close ally of France. An expedition was at once planned against the Dutch Settlements in Guiana; and in September Demerara, Berbice, and Essequibo were captured. A number of the Dutch prisoners were induced to enter the British service, some being formed into a separate corps under command of Lieut.-Colonel McLean of the 2nd Battalion of the 60th, while others were enlisted in our 2nd Battalion itself. In November General Grinfield died and was succeeded, first by General Moncrief, afterwards by Sir Charles Green who arrived in March 1804. Reinforcements, including McLean's corps and the rifle company of our 2nd Battalion, were sent to Dutch Guiana. On May 5 the capitulation of Fort Amsterdam ended the expedition. The rifle company then returned to Barbados. During the rest of the year nothing further of interest occurred, but the mortality among the troops assumed terrible proportions.

In 1805 the arrival of a French squadron caused considerable nervousness in the West Indies, which only passed away on the arrival of Lord Nelson early in June.

On November 7 the British sloop of war *El Orquixo*

foundered off the coast of Jamaica. Out of 136 persons on board, including a detachment of our 6th Battalion, no less than 131 perished.

In 1806 the 3rd Battalion was sent to Guernsey to recruit; and returned to Barbados in March 1807, when it was further strengthened by between two and three hundred volunteers from the 2nd Battalion, and mustered a total of 1300 men. On February 1 of this year the six battalions of the Regiment contained 5687 effectives.

In September 1807 Denmark was added to the list of our foes. In November an expedition of 2500 men, including our 3rd Battalion, 650 strong, under Lieut.-Colonel Elliot, was despatched against the Danish islands of St. Croix, St. Thomas, and St. John's—commonly called The Saints—which surrendered without resistance.

Within eleven months of its return to the West Indies the 3rd Battalion lost 200 men from fever.

After handing over to the 3rd Battalion the large draft already mentioned, the headquarters of the 2nd Battalion was sent for recruiting purposes to Jersey, whence on September 10, 1808, it embarked for Plymouth, with a view to joining the division under Lieut.-General Sir David Baird, who was about to reinforce the army of Sir John Moore in Spain. The battalion was still extremely weak, consisting, as it did, of but 1 field officer, 4 captains, 6 subalterns, 4 staff and 270 non-commissioned officers and men. On October 4 it sailed from Plymouth largely composed of French prisoners of war. On November 3 the battalion reached Coruña, where it disembarked; but when Baird advanced to join Moore was left under Lieut.-Colonel Codd to garrison the town.

In January 1809 Sir John, who had effected his object of making a diversion in favour of the Spaniards and checking the French advance on Portugal, brought his army back to Coruña, and lost his life in the subsequent repulse of a French attack. The 2nd Battalion took part in the battle. Ensign MacArthur, who was present with the battalion, stated in after years that it was in reserve. Some think that by this expression he indicated the division of General Edward Paget which was officially known as 'the Reserve.'

After the action the battalion embarked for England with the rest of the army. It returned from Coruña with 1 field officer, 4 captains, 8 subalterns, 4 staff, 31 sergeants, 20 drummers, and 212 rank and file, and was quartered in Guernsey. By October it had been recruited up to 64 sergeants, 22 drummers, and 1200 rank and file. During the following month it embarked for Barbados.

Lieut.-General Sir George Beckwith was now in command of the Windward Islands. Towards the end of January 1809 he had resolved to make the customary attack upon the island of Martinique. His force was organised in two divisions: the first commanded by Major-General Sir George Prevost, Colonel-Commandant of the 60th; the second, by General F. Maitland. The army comprised nearly 11,000 officers and men, including our 3rd and 4th Battalions, together 700 strong, which formed part of the reserve brigade (commanded by Brigadier-General Nicholson) of Prevost's division.

The expedition sailed from Barbados on January 28 and landed in Martinique two days later. Sir George Prevost's division disembarked unopposed at Bay Robert on the windward coast, and despite difficulties of country, effected an immediate night march of seven miles. Before daybreak of the 31st it occupied a position on the banks of the Grand Lezard River. Maitland's division sailed round and disembarked at St. Luce. On February 1 the enemy's positions at Morne Bruno and Souriry were carried by Sir George Prevost. During the following days further progress was made by both divisions. On the 19th the flank companies of our 4th Battalion received the praise of the General Officer Commanding for their conduct when engaged two days previously with the enemy's picquets close to the Bouillé redoubt. Fort Dessaix surrendered to the 3rd Battalion under Lieut.-Colonel Mackie.

On the 19th Beckwith opened fire upon Fort Bourbon with fourteen heavy guns and twenty-eight mortars, &c. The rain was incessant, but the bombardment continued until 9 A.M. on the 24th, when the enemy surrendered the island. The losses of the British troops in this expedition were comparatively slight.

During the month of April a French squadron, consisting of three sail of the line and two frigates from L'Orient, took refuge at the islands bearing the name of 'The Saints,' where they were blockaded by Admiral Cochrane. Sir George Beckwith at once sent General Maitland with a force of 2800 men, including the 3rd Battalion of the 60th and the flank companies of the 4th Battalion, to reduce the French ports in the islands. On the morning of the 14th in pursuance of his instructions Maitland landed his force with slight opposition at a little bay termed Joli Bois. Advancing inland, the enemy was found posted on Mount Russell, a hill 800 feet high, immediately on Maitland's right. The rifle companies of our 3rd and 4th Battalions were directed to dislodge him.

The ascent was precipitous, at an angle of fifty degrees. Happily the cliff was covered with scrub and prickly pear by which the riflemen pulled themselves up, and not only dislodged the enemy, but inflicted considerable loss upon him. The following General Order was published:

> 'Major-General Maitland has the greatest pleasure in acknowledging the service rendered yesterday (14th) by the three rifle companies of the 3rd and 4th Battalions of the 60th Regiment in dislodging the enemy from Mount Russell. He thanks the captains of those companies—Captains Lupton and Dolling—for their gallant behaviour and great exertions which the Major-General will not fail to report properly to the Commander of the Forces.'

On capturing the heights the whole of the British force occupied Mount Russell, whence a commanding view was obtained of the enemy's camp and the harbour containing the French squadron; but further advance was checked by the fort on the Ile de Cabret which flanked the British left. A battery having, however, been constructed by an officer of Engineers, two eight-inch howitzers were landed, and before 6 P.M. opened fire on the enemy's ships, which evacuated the harbour shortly after dark.

The rifle companies were now detached, with some others under Lieut.-Colonel Prevost, to clear the enemy from a position which commanded the landing point at Ance Vanovre where Maitland, who had re-embarked the bulk of his force, intended to land it again. Prevost achieved his mission, the troops landed, and mortar batteries were erected.

The enemy was posted on Middle Ridge between Forts Napoléon and Moselle. From this he was driven on the night of the 16th by three companies of West Indian Regiments. Next morning the French attempted to retake the position, but were gallantly repulsed by our black troops, supported by the York Rangers and the rifle companies of the 60th. The British loss amounted to about 30 killed and wounded. As a result of this defeat, at midday the French garrison of between 700 and 800 men surrendered. Thirty-four guns were captured. The British loss during the three days' fighting amounted to 6 officers and men killed and 67 wounded. The 3rd Battalion of the 60th lost its captain, Dolling, who fell from a precipice, and one man killed. Lieutenant Van Koning and 16 men were wounded.

Thus ended a brilliant little enterprise admirably combined and conducted by the naval and military forces, although it is only fair to say that the bulk of the French garrison consisted of conscripts. The services of the rifle companies were again acknowledged.

'"Saints," April 17, 1809.

'M. General Maitland again with pleasure acknowledges the bravery and forwardness of the rifle companies of the 60th Regiment under the command of Captains Lupton and St. Germain.'

Before the end of May the force under Sir George Beckwith consisted of 16,000 bayonets distributed among the various islands and in Dutch Guiana. The 3rd Battalion of the 60th, 841 strong, formed part of the garrison of Martinique; the 4th Battalion, 414 strong, was at Antigua. During this month Beckwith issued an order for the supply to the rifle company of the 4th Battalion of new rifles corresponding in length and calibre to those used by the 3rd Battalion.

In November Sir George Beckwith received orders from the Secretary of State in England to attack the French island of Guadeloupe. A battalion of the 60th, 1200 strong, was promised him and, in view of the naval blockade, it was hoped that the operation would be a short one. Beckwith, however, was not very confident, for he knew that our blockade had entirely failed to prevent the arrival of stores and reinforcements, and also that a large number of the black population had been enrolled and armed.

In a despatch, dated December 26, the General mentions the arrival of the 2nd Battalion of the 60th at Barbados.

Pursuant to his orders Sir G. Beckwith made arrangements for an attack upon Guadeloupe. His force of over 6000 men, organised in five brigades, in the second of which was a battalion of grenadiers, 300 strong (including those of our 2nd Battalion), landed in Guadeloupe between January 20 and 28, 1810.

The enterprise was crowned with success. Port Bellair surrendered after a sharp engagement, regarding which, we regret to say, the General Officer Commanding the 2nd division reported :

> 'All troops behaved well except the grenadier companies of the 2nd and 4th Battalions of the 60th composed of foreigners who, notwithstanding every exertion of their gallant officers, did not conduct themselves to their own credit or my satisfaction.'

Writing a few days later to the Secretary of State, Sir G. Beckwith raises the whole question and drives home the moral.

> 'I have no confidence,' he says, 'in troops recently arrived from England in a motley mixture. The flank companies of the 2nd Battalion of the 60th, and the grenadiers of the 4th Battalion behaved shamefully in the face of the enemy. I repeat my request that certain corps in this army be no longer loaded with men polluted by civil crimes, congregated with French prisoners and deserters.'

In truth, had the authorities set their minds on the destruction of a Regiment, ingenuity would be at a loss to discover any method better than that which they were adopting. If the 60th still exists and has earned a reputation for good conduct, it is indebted, not in the smallest degree to the War Office, but to the spirit and tradition of Henry Bouquet and Augustin Prevost, which during all those dark years passed in the West Indies were never entirely extinguished.

With the capture of St. Martin's and St. Eustatius during the same month, the power of France in the Windward Islands became extinct.

In the distribution of Beckwith's army, which exceeded 16,000 rank and file, during March 1810, we find the 2nd Battalion of the 60th, 1074 strong, at Barbados; the 3rd Battalion (746) at Martinique; and the 4th with one wing (390) at Dominica, and the other of equal strength at Ste. Lucia. The wing at Prince Rupert's, Dominica, suffered terribly from fever during the summer, and lost no less than 150 men. At one time the Lieut.-Colonel, one Captain, and the Paymaster only were fit for duty. At another, but two officers were off the sick list.

At the end of the year the cadre of the 4th Battalion was sent home to England and quartered at Lymington, to be recruited. In the spring of 1812 it was ordered to Quebec, but the orders were subsequently cancelled and the battalion was sent to Dominica, where it arrived on June 10, over 1000 strong, but composed of foreigners, many of them being Frenchmen from the prison ships, and the rest deserters from the enemy's armies.

During 1811 a partial insurrection of the black population of Martinique took place, and it was to our 3rd Battalion that the white inhabitants were principally indebted for their safety. The brigade orders dated St. Pierre, September 19, 1811, contained the following paragraph, showing that in spite of all difficulties the battalion maintained a good reputation under the command of a competent officer:

'Major-General Wale begs Lieutenant-Colonel Mackie to accept his best thanks for his vigilance in the late commotion at St. Pierre and for the zeal he manifested on that occasion in supporting the civil authority; thereby giving security to the town. The Major-General also begs Lieutenant-Colonel Mackie to express his approbation to Captain Hinckleday, Lieutenants Sergeant and Fontane and the detachment who voluntarily offered their services and went in pursuit of the banditti; as also his thanks to the officers, N.C.O.s and privates of the 60th regiment for their general and peaceable conduct, which the Major-General is happy to say has been repeatedly noticed to him by the magistrates and most respectable inhabitants of St. Pierre, who seem to place great confidence in this regiment for the maintenance of good order in this populous, and at one time turbulent town.'

Monsieur Valmonier, the Procureur-Général du Roi, also addressed Colonel Mackie on behalf of the white population in terms of gratitude. Three years later, when Mackie left the island, he was presented by the inhabitants with a piece of plate of the value of 1000 dollars. At the same time the General Officer Commanding complimented the Colonel on the efficiency of his battalion in the field, and its good conduct in quarters. The General had 'invariably remarked the happy art by which he [Mackie] had obtained these essential objects with the least possible coercion or severity.'

The distribution of troops on April 6, 1812, shows seven companies of the 2nd Battalion of the 60th (627 rank and file) at Demerara and Berbice, and the remaining three companies (257) at Tobago. Three companies of the 3rd Battalion (273) were at Ste. Lucia; four companies (414) at Martinique; and three (242) at Dominica.

During the summer of this year the United States of America declared war against Great Britain, mainly in consequence of the British Orders in Council which interfered with American trade, and of England's insistence on a right to search American vessels for deserters from our own navy. The British Orders in Council had, in point of fact, been revoked a few days before the American declaration of war. But, contrary to English expectation, peace did not result, and the war in fact lasted for two and a half years.

In this contest the first experiences of the 60th were in the capacity of marines. A detachment of the 4th Battalion was on board H.M. Treasury ship, *Emma*, when attacked by an American privateer of superior force. The privateer was beaten off after a hard fight.

In November Major-General Carmichael at Demerara twice in general orders paid a compliment to the conduct of Lieutenant Adair and a detachment of the 2nd Battalion while on board ship. On the second occasion H.M.S. *Foam* had been aground on the bar, and a superior force of the enemy threatened an attack, but thought better of it. The vessel was then brought safely into the road.

During the same month Sir G. Beckwith, writing to the Secretary of State, made a bitter complaint that the troops under his command were no longer a British army, the majority of the Europeans being French deserters.

In 1814 the war with France for the moment ended. That with the United States continued. During the summer Sir G. Beckwith went home, and was succeeded by Sir James Leith, who had served as a general of division under Wellington.

An attack made by a British force upon New Orleans near the mouth of the Mississippi, in January 1815, was repulsed with great loss. No part of the 60th was present. The British troops then landed in Alabama and captured Fort Mobile. Leith prepared to reinforce them with a brigade which included the 3rd Battalion of the 60th, but a delay occurred and the conclusion of peace rendered the reinforcement unnecessary.

CHAPTER XIII

THE FIRST, FOURTH, AND SIXTH BATTALIONS—WAR WITH SPAIN—THE FOURTH BATTALION AT THE CAPE OF GOOD HOPE—AT BARBADOS—THE FIRST BATTALION SAILS FOR SOUTH AFRICA—SEVENTH AND EIGHTH BATTALIONS RAISED

We must now go back a little to say a word about those battalions of our Regiment, viz. the 1st, 4th and 6th, which, at the rupture of the Peace of Amiens, were quartered in Jamaica.

On July 1, 1803, the 1st Battalion (strength 400, of whom 50 were sick) was distributed at Port Maria and other small posts on the north side of the island.

The 4th Battalion (600 strong, of whom 70 were sick) had six companies at Kingston and four at Spanish Town.

The 6th Battalion (with the same strength as the 4th and with the same number of men sick) was at Up Park Camp.

At about this period the total force in Jamaica under command of Lieut.-General Nugent amounted to nearly 5000 men. By January 1805 it had sensibly dwindled, in spite of the General's attempt to enlist foreigners, and re-engage time-expired men of the 60th. During the following month he made a proposal to draft the men of the 4th Battalion (by this time less than 400 strong) into the other two, and to send the headquarters to England.

Just at this time General Nugent was informed by the Secretary of State that Spain had declared war against Great Britain. The circumstances leading to this unfortunate event require a word of explanation. For some years past an offensive and defensive alliance had existed between Spain and France, and the rupture of the Peace of Amiens should consequently have led to immediate hostilities between Spain and England. Spain, however, had no personal quarrel against Britain, and a compromise was arranged between the allies under the terms of which Spain was allowed to remain at peace, provided that she furnished France with specie at the rate of not much less than three millions a year.

In point of fact no arrangement could have been more convenient to France or more detrimental to Great Britain. Other acts showed that, however little desirous Spain may have been for war with England, she was nothing but a tool in the hands of her powerful neighbour. After remonstrating in vain against various equivocal practices, the British Ambassador demanded his passports and left Madrid on November 10, 1804.

Even then hostilities did not follow until news of a high-handed action, which had taken place two months previously, reached Madrid. The British Government had given orders to stop all ships coming from South America with specie, knowing that such specie formed the sinews of war to our enemy, France. In consequence of this order Admiral Graham Moore (brother to Sir John), cruising near Cadiz with a squadron of four frigates, demanded the surrender of a Spanish squadron of equal numerical strength. The Spanish commander very properly refused compliance. An action took place; one of the Spanish ships blew up; the other three were captured and found to contain treasure to the value of some two million pounds. It was not unnatural that, on December 12, when the

Spanish Government heard of this proceeding, it at once issued a declaration of war.

So far as Jamaica was concerned, the upshot of this quarrel was that British Honduras was in imminent danger of Spanish attack; and, as General Barrow, the Commandant, was out of health, Lieut.-Colonel Gabriel Gordon of the 60th was sent to defend British interests in that quarter.

During the spring considerable alarm was occasioned in Jamaica by the appearance of a French fleet which had eluded the British blockade of Rochfort. Martial law was proclaimed and all measures were taken for defence, but nothing actually happened beyond the attack on Dominica already noticed.

In July the garrison of Jamaica totalled 4730 of all ranks. The 1st and 6th Battalions of the 60th amounted each to 750; numbers which, in February 1806, were reduced to 599 N.C.O.s and men in the case of the 1st Battalion, and to 663 in that of the 6th.

In February Sir Eyre Coote succeeded Nugent, and the military history of the island resolves itself into a gloomy record of mortality.

In 1808 came news of the Spanish revolt against Napoleon and his projects. Offers of British assistance to Cuba and the other Spanish colonies of Central America were forwarded to the Governors. A small British force, including the flank companies of our 1st and 6th Battalions, was made to embark for Cuba, but its services were not required. It was perhaps fortunate, for General Carmichael thus reports upon the selected contingent:

'The 18th broke down hopelessly from fatigue in a night march of 13 miles on an easy road. Active service in this island would break down three-fourths of the men, owing to salt food and excessive drinking. The 2nd W.I.R. has long had exclusive duty in guarding prisoners of war. In one case of a European regiment, the sergeant and nearly the whole of his guard of 18 men were carried, in a few hours, from the guard-room to an hospital.'

The mortality of two regiments quartered on the south side of Jamaica amounted, in a given time, to twenty per cent.; but whether the 60th were better looked after, or better inured to the climate, their death-roll during the same period amounted only to one per cent.

In accordance with General Nugent's arrangements, the 4th Battalion embarked in the *Suffolk* and the *Perseverance*, and landed at Cowes, in the Isle of Wight, on September 30, 1805. In July 1806 the battalion sailed from Portsmouth for the Cape of Good Hope.

In January of this year an expedition under Lieut.-General Sir David Baird had captured Cape Town from the Dutch, after an action in which the Waldeck Rousseau Regiment, composed of Austrian and Hungarian prisoners who had been sold to the Dutch, had behaved without distinction. On the arrival of the skeleton of the 4th Battalion a large proportion of the Waldeck regiment was, in accordance with the bad practice of the time, drafted into the 60th, which was quartered at Cape Town and Simon's Bay, and sometimes encamped at Rondebosch.

Colonel Robert Craufurd touched at Cape Town on his way to South America, and his staff officer speaks of our 4th Battalion as being 'disaffected,' a fact which under the circumstances is hardly a matter for wonder.

In March 1808 the battalion, under Lieut.-Colonel Austin, embarked at the Cape for Barbados, where it arrived in May.

The 1st Battalion, having handed over to other battalions all its efficient men, quitted Maroontown in Jamaica and was sent home in 1810 to Hilsea, where its Colonel-Commandant, General Whetham (at that time Lieut.-Governor of Portsmouth), filled up its ranks from the prisoners of war (excepting those of French birth) to be found in the neighbourhood. From Hilsea the battalion went on to Cowes whence, on May 30, 1811, it embarked (1000 strong) for the Cape of Good Hope. On arrival it was stationed at Grahamstown, on the frontier.

In June 1809 Major-General Carmichael embarked a force of 1300 men (including 200 of the 6th Battalion of the 60th) and in due course landed in Hayti, near the city of San Domingo. The remnant of the French garrison surrendered without resistance.

The next few years were marked by no events of interest. On December 25, 1811, the 6th Battalion had been recruited to a strength of 38 Officers and 1031 N.C.O.s and men. In April 1812 a detachment of 3 Officers and 66 N.C.O.s and men of the 6th Battalion was sent to Curaçoa. In August the total strength of the battalion, exclusive of officers, was 712.

A 7th and an 8th Battalion, whose short history will be touched upon in another volume, were raised in 1813, and on December 25, 1815, the strength of the various battalions of the Regiment was as follows:

1st Battalion	956	rank and file
2nd "	981	"
3rd "	788	"
4th "	876	"
5th "	216	"
6th "	607	"
7th "	776	"
8th "	496	"
Total	5696	"

Previous to this, the military authorities had resolved to order uniformity of dress throughout the Regiment. Hitherto the 5th, 7th and 8th and the rifle companies of other battalions had worn green, while the remainder were still dressed in red. By an order of the Commander-in-Chief, H.R.H. the Duke of York, issued in October 1814, the whole Regiment was directed to wear the green uniform with red facings.

APPENDICES

APPENDIX I

THE REGIMENTAL MOTTO

On October 11, 1824, the following letter was written by the Deputy Adjutant-General, Horse Guards, to the Officer Commanding the battalions of our Regiment:

'Sir,

'I have the honour to acquaint you, by direction of the Commander-in-Chief, that His Majesty has been pleased to permit the 60th Regiment, "The Duke of York's Own Rifle Corps," to resume the motto "Celer et Audax," which was formerly worn by the Regiment in commemoration of its distinguished bravery whilst employed with the British Army in North America, under Major-General Wolfe, in the year 1759.

'I have the honour, etc.,

(signed) 'J. McDonald, D.A.G.'

This letter was written in response to a memorial requesting the resumption of the motto. As to the circumstances under which the motto had been originally conferred, we have only two sources of information.

Major Patrick Murray states that when, in July 1759, General Wolfe had taken up a position on the left bank of the St. Lawrence near the Falls of Montmorenci, the two grenadier companies of our 2nd and 3rd Battalions cleared the country below the falls from the annoyance of the Canadians and Indians, led on by a gallant priest[1] who was killed in one of the skirmishes.

[1] This no doubt was M. Robineau de Port Neuf, Curé of St. Joachim, who with thirty parishioners took post in a large stone house at Château Richer hard by, and for a time held the English at bay. The party was at length enticed outside, surrounded, and, horrible to relate, put to death, and scalped by Montgomery's Rangers, disguised as Indians.

'The History of Services of the 1st Battalion,' written about 1824, copies Murray's statement, and goes on to say, 'In executing this service these Companies displayed so much alertness and intrepidity on all occasions, and General Wolfe being so much pleased with their spirited conduct, conferred upon them the appropriate motto "Celer et Audax."'

The other source of information is as follows: Among those present at the Regimental Dinner in 1869 was Captain William Pearce, who stated that, in July 1824, when the 1st Battalion was waiting at Quebec to embark for England, he had wandered into the Cathedral where he was accosted by a monk who said that he was interested in the 60th owing to the fact that his father had been Sergeant-Major of one of the battalions in 1759, and that during the operations near the Montmorenci Falls the French commander, the Marquis de Montcalm, had constructed a *flèche* or redan, in which he had placed two guns, which brought a flanking fire to bear upon the British and caused great annoyance.

Wolfe directed a detachment to capture the work, but it was repulsed; and the General was standing angry and excited at the failure, when an officer of the Royal Americans came up and said he would take the redan if the General would allow him to have the two grenadier companies of the 2nd and 3rd Battalions of the 60th. Wolfe at first would not listen, but after a little time assented. Thereupon the officer, with two companies, scouted rapidly by hollow ways and through brushwood till he got in rear of the *flèche* unnoticed, and dashed in at the gorge.

The enemy, taken entirely by surprise, was easily driven out of the work; the two guns were spiked, and the Royal American companies returned to their original position, whereupon General Wolfe complimented them handsomely upon their operation and the manner in which it had been carried out, and said that their motto in future should be 'Celer et Audax.'

The monk showed Captain Pearce an old order book which had been kept by his father, the Sergeant-Major. This book contained Wolfe's 'Orders of the Day,' praising the Regiment and giving it its motto. When Captain Pearce came home he brought with him either the original order book or a copy, on production of which the King authorised the resumption of the motto.

This statement of Captain Pearce was heard by Major-General Astley Terry and, no doubt, by others still living who were present at the dinner in question. General Rigaud, who records the fact in a little pamphlet entitled 'Celer et Audax,'[1] was not present, but

[1] Oxford: Pickard Hall & Stacey, printers to the University. 1882.

on hearing the story a few months afterwards, attempted to communicate with Pearce. The latter, however, died early in 1871, but his nephew, Mr. Oakley, in a letter to Rigaud, said he had often heard his uncle give the same account of the origin of the motto. Although the story of Captain Pearce mentions a specific instance for the awarding of the motto, whereas the Regimental record appears to imply that it was bestowed as a result of several exploits, there is no real discrepancy between the two accounts. We may fairly hope that the Regiment earned Wolfe's praise on more than one occasion, and that the capture of the redan was merely the culminating point. As to the exact day on which the episode occurred, one can only guess. In the wild attack of July 31, a redoubt was captured, but it would seem certain, not under the circumstances described by Captain Pearce; nor, in view of the subsequent repulse, was Wolfe likely to have indulged in words of commendation. Ten days previously the 3rd Battalion and the two grenadier companies, together with the 15th and 48th Regiments, had made a successful raid on the left bank of the St. Lawrence, twelve miles above the town: on this occasion a work was taken, and it is not impossible that the unknown officer who asked leave to take the *flèche* with the grenadier companies, was Augustin Prevost; but Wolfe was not present in person, and the story of the monk does not exactly fit in with the circumstances of the case, although of course it is quite possible that lapse of time had affected the literal accuracy of the story. Application to be allowed to attack the *flèche* may have been made by Carlton, the Brigadier, and it is possible that it was upon his report that Wolfe made the exclamation.

A third possible occasion was August 9, on which day General Orders state that 'two Grenadier Companies of ye Royal American Battallions are to embark in four flat-bottomed boats to fall down with ye tide and escort ye General at 6 A.M. to St. Joachim.' What happened on this occasion is not recorded; but supposing the *flèche* to have been captured on that day, Pearce's story, that it happened under Wolfe's eyes, and the Regimental record that it happened down stream, are in harmony.

The reason for the discontinuance of the motto can also be only matter for surmise. It may have been intended only for the Grenadier Companies; it may have been given to the 3rd Battalion, which was disbanded four years afterwards; but the most probable explanation is that its discontinuance was ordered because there was no official authority from the War Office for the Regiment to assume it.

APPENDIX II

MEMOIR OF MAJOR PATRICK MURRAY, WHO SERVED IN THE 60TH FROM 1770 TO 1793

NOTE.—The 'History of the Services of the 1st Battalion' states that the original manuscript, now in the hands of Hamilton J. Bunbury, Esq., was given by Major Murray to the officer commanding the 1st Battalion at Montreal in 1822—presumably Colonel Alexander Andrews,—on whose death it was no doubt handed to his successor, Colonel Bunbury, who seems to have appropriated it.

The original is faulty in punctuation and deficient in capital letters. Words are often omitted. The meaning also in places is obscure. Great care has been taken in the present copy, but it is quite possible that, in one or two paragraphs, what the author intended to convey has been misread.

The first part describes the early history of the regiment, which he may have got at first hand from officers who had been original members of the Royal Americans. His statements were to a great degree copied word for word in the 'History of Services of the 1st Battalion,' except for a few errors by the transcriber. The second part narrates the author's personal experiences in the War of Independence. The memoir ends abruptly and is evidently an unfinished work. Its value consists chiefly in the fact of being the only known contemporary record of work done by a part of the Regiment in the War of Independence.

Words shown between brackets [ ] are not found in the MS.

PART I

'After the disastrous events in the back settlements, particularly the surprise of Logstown and Fort Monongahela on the Ohio, the capture of Colonel Washington and the total defeat of General

Braddock at the Meadows near Fort Du Quesne in the years 1754–1755, a plan was formed, under the patronage of His Royal Highness the Duke of Cumberland, then Commander-in-Chief of the British Army, to raise a body of Troops for Service in North America, which was to be levied chiefly from amongst the German Colonists in the back Settlements of America and to be completed by Levies from Germany and Great Britain.—The original Projector of this scheme is said to have been M. Von Harbot, a Gentleman of the Canton of Bern who brought it forward through the interest which Colonel James Prevost of Geneva, then a Captain in the Prince of Orange's Swiss Guards, had with the Princess of Orange, sister to the Duke of Cumberland.

1754
1755.

'It being contrary to the Act of Settlement to admit Foreign Troops into the British Line, it became necessary to have an Act passed, authorizing 60 officers and 1200 Men, with 20 Engineers, Foreign Protestants, to serve in North America.[1] While the Debates on this subject were Pending 10 New Regiments were raised, in consequence of which delay the Royal Americans were numbered the 62nd., instead of being the 52nd. Regiment and were composed of 4 Battalions as follows :—

1756.

Germans raised in the American Colonies . .	1800
Foreign Protestants raised in Germany . .	1200
Drafted from Regiments in Great Britain . .	200
Drafted from Regiments in Ireland . . .	800
Total . .	4000[2]

'There were also 20 Foreign Engineer Officers but not included in the strength of the Regiment.

'The 1800 Provincials raised in America were Levied under Lord Loudon's Proclamation, his Lordship being then Commander-in-Chief of the Forces in America and was appointed the first Colonel-in-Chief of the Royal Americans the 25th December 1755.—

'The first Battalion was commanded by Colonel Stanwix, the second by Colonel Dusseaux, the third by Colonel Haveland, the fourth by Colonel John Prevost.

[1] The Act specified fifty foreign officers, not sixty. It said nothing about 1200 men.

[2] The writer has found no evidence that any recruits were brought from Europe. Murray's statement refers perhaps to the original intention.

'When in consequence of the capture of the 50th. and 51st. Regiments at Oswego these two Regiments were disbanded the Royal American Regiment became the 60th.

1st. Battalion.

'The 1st. Battalion was distributed after the French were expelled from Canada, on the communication with the Ohio and the Lakes, from Philadelphia to Lake Michigan and Superior. They assisted at the taking of Pittsburg and furnished garrisons to Forts Lyttleton and Bedford. When Colonel Bouquet with the remains of the 42nd. and 77th. scarcely amounted to 500 men lately from the Havanna marched to the relief of Pittsburg, parties of the 1st. Battalion of the 60th. were drawn from the scanty garrisons on the route to act as advance guard, and escort to a convoy of Provisions and stores which it was the object of the expedition to throw into Pittsburg. On the evening of the 5th. August 1763 after a march of 17 miles the advanced guard party, having entered the defile of Bushy Run on the banks of which they intended to encamp was assailed by a heavy fire from the Indians in ambush which killed and wounded the whole except a Sergeant and 2 men who, notwithstanding, resolutely stood their ground, returning the fire till sustained by the Highland Light Infantry. The action at Bushy Run which reflected so much honour on Colonel Bouquet and the troops under his command, is recorded in the Annual Register for 1763. The Indians were dispersed and Pittsburg relieved.

'The following campaign 1764 Brigadier General Bouquet advanced with a corps of Regulars and Provincials as far as the Muskingum where the Indian Tribes submitted, and delivered their captives to Colonel Bouquet. In the same war Colonel Bradstreet advanced by the Lakes as far as Sandusky where he received the submission of Pondiac whose forces had in 1763 destroyed the small garrisons of Toronto, Presqu'isle, Le Beuf, Venange, Sandusky, Miamis, St. Joseph and even by stratagem Michilimakinac where 2 Companies of the 1st. Battalion were massacred except Captain Etherington the commanding Officer, and 11 others who were saved by the Missionary Jesuit;[1] and they were escorted to Montreal with the garrison of La Baye St. Joseph by a body of Sakis with their arms which Sergeant Noel[2] refused to deliver up; and the Indians admiring his bravery, took the party under their protection, consisting of Ensign Gorrell and 15 men. The other small garrisons were massacred; but Detroit was gallantly defended by Major Gladwin of the 80th. Light

[1] Father Jonois.—L. B.

[2] *Quære* Null.

Infantry with a garrison composed of part of that corps and of the 1st. Battalion 60th ; Niagara was too strong to be attacked by Indians. A Sergeant of the 60th named Smith, distinguished himself in a remarkable manner in the defence of a schooner with 7 men, against 2 or 300 Indians whom he forced to leap overboard by jumping with a lighted match down the hatch-way and threatening to blow the vessel up ; when he coolly hoisted sail and entered the harbour, himself and party being all over blood from the wounds they received from skalping knives. The Indians fled to an Island called " fighting island " to this day.

'A bloody skirmish took place also between a detachment of the garrison commanded by Captain Dalzel aid de Camp to General Amherst, commander-in-chief, where Captain Dalzel and others were killed. Three companies of the 1st. Battalion had been detached from New York to Carolina and Georgia which returned to Canada in 1766. The Regiment was reduced at the peace 1763 to 2 Battalions of 450 men each, 9 Companies ; 2nd. under The Hon: M. G. Murray [1] and Prevost 1st.

2nd. Battalion.

'The 2nd. Battalion was sent to Nova Scotia in 1757 where its first action was an affair with the French who were worsted. [In] 1758, it was at the siege of Louisbourg which was taken, and in 1759 with General Wolfe at Quebec, leaving 200 men to replace the Grenadiers of the 22nd, 40th. and 45th. which with those of the 2nd. Battalion 60th. formed the Corps called the Louisburg Grenadiers, under Lieutenant Colonel Alexander Murray. The following Corps composed the Army under Major General James Wolfe.

'Brigadiers General Hon: George Townshend, Hon: Robert Moncton, Hon: James Murray.

'Quarter Master General Colonel Guy Carleton.

'Adjutant General Major Isaac Barré.

'Chief Engineer Major Mackellar.

'3 Companies of Royal Artillery, and the following Regiments of foot, 15th., 28th., 35th., 43rd., 47th., 48th., 58th., 2nd. 60th., 3rd. 60th., 78th ; Louisburg Grenadiers ; a Company of Light Infantry from [each regiment in] the Army ; Company of Light Infantry from the garrison of Louisburg ; 6 Companies of New England Rangers ; one Company of New England Carpenters. The Grenadiers of the Army under Colonel Carleton. The Light Infantry, under the Hon: Colonel Howe. The army was occasionally strengthened by the Marines of the fleet under Lieutenant Colonel Hector Boisrand.[2]

[1] J. is Murray's real initial.—L. B.

[2] *Quære* Boisland.—L. B.

'The Hon: Brigadier General Monkton had been appointed Colonel Commandant of the 2nd. Battalion 60th. vice Dusseaux.

'The army landed on the Isle of Orleans, 26th. June 1759.

[On the 9th July] 'the army landed on the north shore below the fall of Montmorency. Monkton's Brigade of 4 Battalions crossed over to Point Levy where after some skirmishing with the Canadians they took post and erected Batteries to bombard and Cannonade the Town. The 2 Grenadier Companies of the 60th. cleared the country below the falls from the annoyance of the Canadians and Indians, led on by a gallant Priest, who was killed in one of the skirmishes.

'On the 31st. July, General Wolfe having concentrated his army crossed the Montmorency below the falls, to attack the French redoubts and intrenchments with the 13 companies of Grenadiers and 200 of the 2nd. Battalion who stood on the left of their Grenadier Company, which happened to be the left of the Grenadiers, when Captain Ochterlony said to Captain Wetterstrom that though his men were not Grenadiers they would be the first to storm the French redoubt. The Grenadiers taking fire at this speech the whole line rushed furiously on without orders ; the first redoubt was abandoned, but the disorderly attack was checked by the heavy fire of the French and the extraordinary steepness of the ascent ; still they persisted as if they alone could beat the whole French army ; happily a violent thunderstorm enabled the officers to withdraw this brave but mistaken body, with the loss of upwards of 400 officers and men killed and wounded ; the 15th. and 78th. being drawn up in good order along the shore, and General Murray's Brigade crossing at low water, General Montcalm wisely preferred relying on the strength of his position to the risk of an encounter with such desperadoes. Captain Ochterlony a gallant officer and accomplished gentleman, was mortally wounded and taken to the French General Hospital where he died. The Grenadiers [were] ordered to join their corps.[1]

'On the memorable 13th. of September, the line consisted of the Louisburg Grenadiers (those of the 2nd. Battalion being the left Company. General Wolfe was killed at their head), ye 47th., 35th., 28th., 58th., 43rd., 78th., 35th., under the Brigadiers Monkton and Murray. The 15th. and 2nd. Battalion 60th. under Brigadier Townshend were *en potence* facing the River St. Charles with Colonel Howe's Light Infantry. The 48th. in 4 Gr[and] Divisions

[1] Utterly wrong as was this affair, contemporary writers say it was the admiration of the whole army.—L. B.

formed the reserve, and the 3rd. Battalion 60th. was at the Foulan to guard the landing. The light company of the 2nd. Battalion had been detached under Lieutenant Mac Alpin who ascended a woody precipice and drove a detachment of 100 men out of the bishop's country house, and routed another detachment of 100 men who came from Sillery to retake it.

'In the Battle General Wolfe was killed and General Monkton supposed to be mortally wounded, but recovered and with the rest of the wounded was sent to New York by sea.

'The French General Montcalm was killed in the action and his 2nd. in command.[1]

'After the French line was broken 2000 chosen men of the French under M. de Bougainville appeared in the rear of the army, but were kept at bay by the 15th., 2nd. Battalion 60th. and the Light Infantry under Colonel Howe. The 2nd. Battalion had about 100 killed and wounded.

'Quebec surrendered 18th. September. General Townshend returned to England and General Murray remained as Governor of Quebec with about 6000 men; some recruits from Europe and the South having joined during the siege. The 2nd. Battalion had its share in the hardships of the winter 59–60, the memorable defence, the Battle of 28th. April, and the pursuit of the French to Montreal, the junction with Generals Amherst and Haviland and the final reduction of Canada when the 2nd. Battalion returned to Quebec with Governor Murray who was appointed to be Colonel Commandant (vice Monkton promoted to an old regiment) and remained there during the rest of the war.

3rd. Battalion.

'The 3rd. Battalion was employed in 56, 57, we think, about the Countries between Albany and the Lakes; they were with General Webb at William Henry, and in 1758 with General Amherst at Louisburg, about which time Brigadier Lawrence became the Colonel Commandant. At the siege of Quebec in 59, they were commanded by Major Augustin Prevost, who was shot through the skull and trepanned with success [and] brought to New York by sea; the Battalion in the meantime [being] commanded by Major Young.[2] Lieutenant Colonel Prevost joined the 3rd Battalion in 1760, conducted them to the siege of Martinique and the Havannah, where they distinguished themselves by repulsing,

[1] M. de Sénézergues.—L. B.

[2] Young was Lieutenant-Colonel of the battalion. The photograph of Prevost facing p. 228 shows where he was trepanned.—L. B.

(being brigaded with the 35th.) a sally from the Moro on the heights of Cavagnos where they were carrying on the Approaches and drove the Spaniards into the sea and took many of them in their B[attalion] in a remarkable action with a Spanish Regiment of Edinburgh Dragoons, and a multitude of mounted militia which the 3rd. Battalion defeated at Guanamacoa, by plunging through a morass and charging and dispersing the troopers. The passage of the breach of the Moro was discovered by Captain Lieutenant Fuser [1] 3rd. Battalion, which being stormed led to the conquest of Cuba.

'After the peace 1763 the 3rd Battalion took possession of West Florida ; and being relieved by the 22nd. and 34th. returned to Europe to be disbanded. The Radeau employed by General Murray in the defence of Quebec was constructed under the direction of Captain Watterstrom of the 2nd. [Battalion] who volunteered with the 3rd. Battalion to [the] West Indies.

4th. Battalion.

'The 4th. Battalion served under Lieutenant Colonel Haldimand on the station on Lake Ontario's communication with Hudson River. At the battle of Tycondaroga Colonel Haldimand commanded the Grenadiers of the Army, and Major Baron Munster led the 4th. Battalion to the attack of the Lines where the army was repulsed with great slaughter ; but the 4th. Battalion stood their ground and persisted with the 42nd. and other corps till recalled by repeated orders.[2] Captain Williamoz was distinguished as a partizan and Engineer, during the campaign of 1759 ; and the 4th. Battalion served in the army under General Prideaux who was killed besieging Niagara by the blunder of the Engineer who threw the parapet behind the trench, which blunder Captain Williamoz was sent for from Oswego to correct ; and the fort was taken after a spirited sally and attempt to relieve the garrison by the French. It surrendered to Colonel Sir Will Johnson. Oswego surrendered also ; General Amherst advanced to Ticondarogo and Crown point upon Lake Champlain [in] 1760. The 4th. Battalion acted with the main army under General Amherst by Oswego and the Lakes. That army descended the St. Lawrence after a slight resistance at La Galette or Oswego, and the loss of 400 men in the rapids, and landed at La China with about 16,000 Regulars and provincials with 1000 Indians under Sir W. Johnson. On the

[1] Fuser, A.D.C. to General Keppel. The words 'A.D.C. to' are erased in the MS. and 3rd Battalion substituted.

[2] Baron Munster is returned in the list of wounded of the 1st Battalion.—L. B

same day General Murray land[ed] at Pointe aux Trembles, having been joined in his progress up the River by Lord Rollo with the 22nd. and 40th., General Haviland moved over Lake Champlain, and after a faint show of resistance, Isle aux Noix being abandoned, that General reached La Prairie on the same day [that] Generals Amherst and Murray landed on the Island of Montreal. The French Governor M. de Vaudreuil capitulated for all Canada, and the French Army which had concentered at Montreal laid down their arms [under promise] not to serve during the war on account of the cruelties they suffered the Indians to commit, and the barbarous practice of making the female captives slaves. The terms to the people were very favourable and mild. The militia were allowed to retire to their homes, delivering up their arms. The Acadians [1] also were favourably treated, but the French army consisting of 8 Battalions and 40 Companies of Colony Troops or Marines were embarked for France with their General Monsieur de Levi, and the Marquis de Vaudreuil etc. Brigadier General Gage was appointed Governor of Montreal and Brigadier Burton of 3 Rivers.[2] The 4th. Battalion was cantoned in the Governments of Montreal and 3 Rivers till late in 1763 when they were embarked for England and disbanded. Colonel Haldimand became Lieutenant Colonel of the 2nd. Battalion.'

'At the peace 1763 General Gage was appointed Commander-in-Chief but General Amherst remained Colonel-in-Chief of the 60th. after his return to Europe.

'General Murray was made Governor of the Provincials and Military Governor of Quebec: Colonel Burton Brigadier of the District.

'Captain, afterwards Major General, Des Barres was employed by the Admiralty in his admirable survey of Nova Scotia, but never did duty in the Regiment after the siege of Quebec. He was gratified with rank in the army after he had ceased to serve in it, and was appointed governor of Cape Breton from which he was removed, for indiscreatly taking upon himself to break the 33rd. Regiment, yet he was a man of superior merit in his proper profession as a surveyor, draftsman and field Engineer.

'Captain Holland was made a Surveyor General of Lower Canada and of the Northern District which he carried on including

[1] I.e., roughly speaking, the inhabitants of Nova Scotia.—L. B.

[2] Les Trois Rivières.—L.B.

New Hampshire and Rhode Island much to his credit. General Murray, or the Duke of Richmond got him into the Council but did not illuminate his mind with the knowledge of the civil arcana.

'Lieutenant Ratzer surveyed the Jerseys, and was afterwards Surveyor General of Jamaica where he died a brevet Major commanding the 1st Battalion.'

'In the year 1770 a company of Light Infantry was added to each Battalion throughout the line.

'In the year 1765 Brigadier General Bouquet died at Pensacola, and Lieutenant Colonel Augustine Prevost succeeded him as Lieutenant Colonel of the 1st. Battalion which he joined from Hambro'[1] in 1766 with about 250 Germans for the 2 Battalions.

'Colonel Haldimand was appointed to succeed Brigadier General Bouquet as Brigadier General.

'The 2nd. Battalion was marched to Montreal in the depth of the winter 1764–1765, to relieve the 28th. which was in a state of mutiny on account of the cruelty of Mr. Walker a justice of peace, in refusing to quarter the families of the soldiers on the inhabitants, after a fire where upwards of 100 houses were burnt. Some persons supposed to belong to the army cut off Walker's ear, dragging him from table while at supper; for which six principal Gentlemen were wrongfully arrested, tried and acquitted, viz: Judge Fraser; Lieutenant Luke la Corne; Captain Disney, Town Major; Lieutenant Evans 28th; Mr. Joseph Howard, Merchant, etc.

'The 2nd. Battalion in the spring 1765 marched to New York, where being ordered to protect a vessel with stamps, the mob pelted the detachment, and no magistrate appearing to read the riot act the Grenadier Company charged and dispersed them without bloodshed. Soon after the Battalion being ordered to the upper country, while the officers were at breakfast previous to embarkation for Albany, the men broke open the prison by way of retaliation, liberated the Debtors and carried off Major Rodgers Governor of Michilimakinac who had been confined for debt. This took place under the feeble command of General Gage. Yet the Battalion was so much esteemed that not the slightest peculation took place during the 6 years they were on that service insomuch that the merchants of New York presented the men with sea stores when they embarked in 1772 for the West Indies. The 10th. Regiment relieved them on the Lakes.

[1] Hamburgh.

'In this year the 26th. relieved the 1st. Battalion (which had been recruited in the Jersies etc.) at Montreal, Crown Point and Tycondaroga and Fort George, and were very injudiciously cantoned previous to embarkation in the Jersies, their native country. General Gage [was much] alarmed at the numbers of leaves granted by Colonel Prevost, [and] though the men all returned faithfully to their companies this mark of mistrust [1] cost this very fine Battallion 110 of their best recruits, who were afterwards replaced at Jamaica by 60 men from the 66th. and 36 from the 36th.[2] who volunteered for the Battalion. It is necessary to observe that Officers and men, considered the Corps as raised for the service of North America, until by the omnipotence of Parliament, Jamaica and the other West India Island were made a part of the Northern Continent of America, though Barbadoes, Grenada and Tobago are almost in sight of South America.[3]

'On General Gage going to England, an act was passed giving Foreign Protestants who had served three years in the Royal American Regiment, the privilege of exercising any employment civil or military in any part of the British Dominions of H.M., Great Britain and Ireland excepted, in virtue of which General Haldimand became Commander-in-Chief in North America, but General Gage soon returned and resumed the Command to carry on a paper war with the Bostonians, for which he was eminently qualified.

'[In the year 1773] the 2nd. Battalion landed 6 Companies at St. Vincent which Brigadier General Dalrymple was invading with the 14th., 29th., 31st., 32nd., 62nd. and 70th. Regiments. Along the shores of the Island Major Etherington after negotiating with the Charibs marched the Light Company to their town, and being a man of consumate [*sic*] address among the lower orders of mankind brought them to a capitulation, for which they granted him a

[1] This is the version given by the 'History of Services of the 1st Battalion.' The MS. runs 'General Gage alarmed at the number of leaves granted by Colonel Prevost, though the men all returned faithfully to their companies, that this mark of mistrust' &c. Some words seem to be left out between 'companies' and 'that.' I suggest they may have been something like the following—'stopped all leave, a measure so much resented' that . . . —L. B.

[2] The number is really illegible in the MS. The 1st Battalion Record says '36th.'—L. B.

[3] Version of the 1st Battalion Record. That of the MS. does not make sense. —L. B.

valuable tract out of their remaining territory. Four Companies of the 2nd. Battalion remained at Antigua, under B[revet] Major Brown.' [1]

PART II

(This contains a narrative of the events in which the author personally took part.)

'The prelude to that great tragedy of errors, the American Rebellion, was now gradually drawing to a crisis. Government raised two additional Battalions for the 60th., partly in Germany, and sent [them] from England to the Floridas; the 3rd Battalion had 6 Companies at St. Augustine, 3 at Pensacolea and one at Port Royal at Jamaica, having escorted Sir John Dalling, their Colonel, to that Island, of which he was Governor, 1776. Pensacola was Head Quarters and Lieutenant Colonel Steele (Stiell) Commandant of the Battalion. Major Van Braam commanded the 6 Companies at St. Augustine.

'The 4th. Battalion had also 6 Companies at St. Augustine. Colonel Prevost, Lieutenant Colonel Fuser and Captain Murray joined from the 1st Battalion at Jamaica, as did Captain Mackintosh for the 3rd. Major Van Braam of the 3rd. Battalion and Captain Wulff [and] Major Glazier of the 4th. also joined from England. Four Companies of the 4th. Battalion were sent to West Florida from England under Captain Christie, as also the Company of the 3rd. from Jamaica.

'The 4 Companies of the 4th. Battalion were taken at Mobile by the Spaniards after a most gallant Defence under Lieutenant Governor Durnford, and joined at Augustine by way of Vera Cruz and the Havanna in the year 1781. The Spaniards durst not assault them, and they marched out at the Breach. Lieutenants Loup and Macdonald distinguished themselves in this defence.

'Lord Amherst sent 3 Companies of the 2nd. Battalion from St. Vincent to assist the new Battalion in their training; but through the management of Major Etherinton the 3 companies consisted of only 82 veterans and 2 Officers, Lieutenant Wickham and Muller; but they were joined in G[arrison] Duty to 5 Companies

[1] This would seem to be the officer into whose arms, according to some accounts, Wolfe fell mortally wounded.—L.B.

of the 14th. under Major Furlong and 3 of the 16th. from Pensacola under Major Graham; and the Recruits were left to the exertions of their own officers, when they could be s[pared] from the works then carrying on under the Engineer Captain Moncrief. who being called to act with the main army, General Howe appointed Lieutenant Colonel Fuser[1] to superintend the fortifications: the place being threatened by the Rebel General Lea who advanced as far as Sunbury for that purpose, but returned when he heard of the prepared state of the Garrison. The American Rebels entered the Province and carried off 1000 head of cattle. A rebel schooner attempted to board a transport brig, which had the Colours of the 3rd. and 4th. Battalions on board with about 80 recruits, and the Ensigns Campbell, Finlay and Porbeck [Probick ?]; but coming under their musketry she was taken and sent in to St. Augustine. Ensign Porbeck [Probick(?)] being Prize Master, the Transport Brig came in next day.

'A skirmish took place across the river St. Mary where it is only 80 yards wide, between Major Graham of the 16th. with 100 men and the Georgian Colonel McIntosh, in which one man of the 2nd. Battalion was killed. Major Graham finding the Enemy was in force moved to the mouth of St. Mary river where H.M. schooner *St. John* and the provincial sloop were at anchor. Some galleys full of men with heavy guns entered Amelia Sound from Cumberland Island, and Lieutenant Grant with H.M. Schooner *St. John* sailed out of St. Mary's sound between Cumberland and Amelia. The troops moved to Amelia and the Provincial sloop was burnt to save her from falling into the hands of the enemy. Major Graham returned by "the Sisters" and "Hester's Bluff," his detachment being sickly. He was relieved by Captain Ross of the 14th. who took post at the Cowford with 100 men of the 14th., 16th. and 2nd. Battalion 60th. A boat race between some Masters of ships in the river took place on the King's Birthday;[2] this attracted several planters, sailors and others to the house of Monsieur Courvoisier, a Swiss, on the left bank where the river at the Cowford is only 1600 yards over.

'The Corps of East Florida Rangers were formed for him.[3] The necessary horses being collected, the detachment proceeded to

[1] C.O. of the 4th Battalion.

[2] June 4.

[3] This seems to be the first occasion on which a part of the 60th was turned into Mounted Infantry. This apparently took place in January or February 1778.—L.B.

the Forts of Nassau 22 miles [distant], consisting of Captain Murray's

	Staff.	C.	S.	S.	D.	R. & F.
Light Company 4th Battalion 60th . .		1	1	2	2	43
Grenadiers of 2nd Battalion 60th . .			1	1	0	16
Light Infantry of 14th and 16th . .				2	1	14
Staff, Lt.-Col. Fuser & Doctor Williams .	2	0	0	0	0	0
	2	1	2	5	3	73
Rangers	1	1	5	0	0	15
	3	2	7	5	3	88
Captain Mackintosh's Detachment Left on St. Mary River		1	1	1	1	50
Total . .	3	3	8	6	4	138

The next day [they] reached St. Mary River at Captain Taylor's house, which being fortified by the detachment was called Fort McIntosh, and Captain McIntosh was left there with fifty men, while the Indians were sweating themselves by way of preparation for war. Colonel Brown with the Rangers proceeded to Satilla where the Americans had burnt the Fort on this side ; they crossed Satilla and the Detachment having constructed a raft crossed the St. Mary and advanced to the burnt Fort. When Colonel Brown came over and brought us some Beef, the Rangers having fallen in with a herd of cattle, a raft was constructed with the burnt wood Palisades of the Fort which were of Cedar, and the Detachment occupied a space between two Rivulets with high banks, and an Indian tumulus in the center being converted into a redoubt. The brush-wood was formed into a line Breasthigh flanked by the redoubt, and the sandy soil being thrown up, the work had the appearance of an entrenched camp. The piled brushwood at the least prevented the rifle men from taking aim, and would easily be cleared for a charge.

'The Rangers found a store with 30 tierces of rice, 7 miles from our Camp, on a creek running into the sea. Captain Murray with 50 men brought off in the men's knapsacs and haversacs 6 tierces of rice and dispatched a ranger to Captain Squires of the Otter, who with his boats took off the remainder. On the return to the Camp a report was made by Colonel Brown, that the Indians had discovered the American Fort 26 miles distant, that the Indians had taken 52 horses, the garrison having retired into the Fort, and

Colonel Brown had fallen back 6 miles, only 17 Indians remaining with him, the rest having gone off with the horses. Upon this Colonel Fuser marched a little before sunset, leaving Serjeant Wools of the 14th. with a party to protect the Camp. The detachment halted at Cantys, 7 miles, and at midnight another ranger brought word that the Americans had dispatched their only 2 remaining horsemen to Fort Barrington on the Altamaha which was 56 miles distant. Colonel Fuser immediately proceeded, and at sunrise joined Colonel Browne, when the whole proceeded to within sight of the Fort, the Indians setting fire to the Woods, and the Rangers moving on the road to the Altamaha. Colonel Fuser and Captain Murray reconnoitred the Fort, and found that a Pond whence a rivulet discharges into the creek between which and the Satilla the Fort was seated, had a winding course through a woody dell into the creek,[1] so that by cutting the brush-wood, a kind of boyau was made by which the troops came within about 300 yards of the Fort which evidently was not defensible. It appeared also that they had no water within the Fort, as one of the Indians wounded with his tomahawk an American who was filling the buckets with water from the creek. A parley was beat opposite the Fort and at last Captain Wynne of the South Carolina Rangers, sent an officer in exchange for Ensign Campbell; and being convinced by our manner that our detachment was the advanced guard of Brigadier General Prevost's army, signed a capitulation, by which he was to deliver the Fort at Sunset, the garrison to lay down their arms and to be escorted to the Altamaha, protected from the Indians. Accordingly Captain Murray with his Light Company drew up opposite the gate which was barricaded, when two Rangers galloped up and whispered him that 300 men were within 4 miles; where, being fired at by 4 of our rangers, they encamped, not chusing at that hour to venture into the swamp where 2 rangers remained to watch them. The men were instantly ordered to spring upon the parapet, which they could easily do, the ditch not being half excavated, and the palisade placed injudiciously inside the parapet with here and there a log to enable the men to fire over it. Captain Wynne immediately opened the gate dreading the bayonets, when 52 Carolina riflemen and 23 of the 1st. Georgian Regiment laid down their arms, and were marched off escorted by Captain Murray and 20 picked Light Infantry. When they came to the swamp where the 300 men encamped they were sent to join them and being

[1] *Sic*, in the original. The real meaning, of course, must be that the rivulet from the pond had a winding course.—L.B.

without arms and showing the Capitulation, by which they were not to serve till exchanged,—it was their interest to magnify our force,—the whole of the Americans broke up their Camp and spreading the report when they reached Fort Barrington [that] they were retreating before General Prevost's army the settlements all broke up as far as Sapelo, Loyalists excepted. In the meantime the escort returned to the fort, and soon after a ranger acquainted Colonel Fuser of the retrograde movement of the rebels. At day-break the rifles were distributed to the Indians ; [we] destroyed the musquets and set fire to the Fort. Two Americans badly wounded were dressed by our Surgeon, and a few Rangers being left to observe with some Indians to spread the alarm, the detachment marched back in one day 26 miles to our Camp with 2 American Officers and 2 men as hostages. The detachment remained 10 days at this Camp, during which time the Rangers drove near 2000 head of Cattle over Satilla and St. Mary's, 1800 of which were brought over St. John River and sold by Government or Tonyn to Messrs. Mackenzie, and Pontio Sanchez, for 25 shillings per head, who sold the Beef at 3d. per pound in the public market. The detachment entered St. Augustine the 5th March without the loss of a man except one Indian killed.

'It is also remarkable that the regulars did not fire a shot and only loaded once, [viz.] when they had taken possession of the Fort.

'In April or May Major Baker's Georgian Dragoons made an incursion into the Province, drove our Rangers and Indians as far as St. John's River and had a scirmish [*sic*] with the Black Creek Factor, in which 2 Indians were killed. Major Marc Prevost of the 2nd. Battalion, with Captain Graham of the 16th. Light Infantry, Captain Wulf of the Grenadiers 4th Battalion, and a detachment of the 2nd. and 3rd. Battalions—the whole did not exceed 100 men—crossed at the Cowford and advanced 11 miles to Rolfe's Saw-Mill, when in the night a Ranger brought the Major one of their horses' ears as a proof of where they were. At day-break they were found at Thomas's Swamp when the Indians began to skirmish with them as they were about to mount, and kept a bushy run of water between them, giving way until at Broadday the Major with his 3 Columns appeared advancing with fixed Bayonets, on which they fled to their horses and attempted to escape through a miry passage in the discharge of Thomas's Swamp but were fired upon by the Indians with Rifles, and attacked by the troops. About 30 of them escaped with Major Baker, about 40 with Captain Fen surrendered; the remainder plunged into the Swamp leaving the Horses; and what became of them is not known. The prisoners were all put to death except 16, among whom was Captain Fen, saved with difficulty by Major Prévost and the Troops from the fury of the Black Creek Factor.

'The American Galleys withdrew to Sapelo, or Sunbury, and they left 3 or 400 men at Frederica under Colonel Elbert.

'An expedition was prepared under Major Glazier; and the Light Company of the 4th. Battalion had advanced to Rolfe's Mill. Next day Captain McDonald came up with the Grenadiers of the 3rd Battalion, when Lieutenant Mowbray of the Navy reported to the Major, that the *Galatea's* Tender, Lieutenant Cranston, and the hired sloop *Rebecca* [were] being despatched to bridle them, while the troops were to attack the Americans on the island. But unfortunately the two vessels grounded in the inland passage called Raccoongut, and the ebb laid their decks open to musquetry which Colonel Elbert took advantage of, and the crews taking to their boats the vessels fell into the hands of the enemy. Thus the expedition proved abortive.

'The *Perseus*, *Daphne* and *Galatea* cruised on the Coast, and brought several prizes into Augustine and St. Mary's; among others the chevalier de Bretigny, 16 of his officers and 200 of his men brought from France in Merchant ships to serve the Congress; but there being no declaration of war with France, nor treaty of Alliance between France and America, these officers were refused their Parole, and remained prisoners in the State house.

'In the course of this year a body of Loyalists from the Forts of Broad River and Saluda, the remains and descendants of the Palatines, forced their way through Carolina and Georgia; and after many of them had been driven back, reached Florida to the number of 260 men with 150 horses. They were formed into 2 troops of Rifle Dragoons of 40, and 4 Companies of Infantry of 45 men each, choosing their officers; they were called the South Carolina Royalists, and [chose] the Superintendent of Loyalists, Colonel Inglis for their Colonel, who never joined them.

'They were under the direction of Major M. Prevost. About this [time] the 5 companies of the 14th. departed for New York and the American General Robert Howe with about 3000 men, of whom 1200 were high up on St. Mary's under Brigadier General Scriven [approached]. General Prevost advanced to St. John's

River. Colonel Brown pushed on to St. Mary's with the Indians and Rangers, and took post at Captain Taylor's house where [he] was surrounded by Scriven, but contrived to get into the cabbage Swamp where they subsisted on Palmeto Fern roots, without the Americans being able to ferret them out. Major Prevost was sent forward to their support with the Grenadiers of the 2nd. Battalion, [the] Light Companies of the 16th. and 4th. Battalion, and the South Carolina Royalists. He took post at Aligator bridge 22 miles from the Cowford with about 450 men. The enemy took a position at Niel Rain's 17 miles from Aligator Bridge. Major Prevost sent forward Major Graham of the 16th. with 200 men with some of Mc Girth's rangers towards the cabbage Swamp to relieve Colonel Brown. The Rangers scouted to the right and fell in with a Column of the enemy who fired upon their advanced men and then changed the direction of their route. Major Graham continued his march until joined by Colonel Brown, when the whole encamped. The morning after Major Graham's advance [there] came into Major Prevost an inhabitant who had escaped from Nassau Bluff 17 miles to our right, and fearing to cross Mill's Swamp in the night, he informed the Major that Colonel White was there with 90 of the 4th. Georgia, and Captain Nash with 14 Dragoons. Major Prevost dispatched Captain Murray with 20 of his company, 20 of the 16th. Light Infantry, and Captain York of the South Carolina Dragoons. As soon as the detachment had got through Mills and Thomas's swamp, Captain Murray sent forward Captain York to occupy the 2 ends of a Pond which formed a half moon projecting outwards; the 2 Points closing in upon the Bluff; but Captain York thought it prudent to keep his men together and took post at the farther end, [upon] which Colonel White judged it full time for him to get over Nassau; and Captain Murray with the mounted [men] pursued; but the last boat load of his rear went down the stream leaving a doctor and his 2 daughters who had joined the enemy, but as the 2 families of Royalists who had been prisoners at the bluff declared Colonel White had used them well, Captain Murray suffered the doctor and his family to depart with their baggage, not thinking it right to allow them to remain in the province. Captain Murray, Lieutenant Calderwood 16th., Campbell 4th. Battalion and Captain York, South Carolina Royalists, made a hearty dinner at the Bluff upon roast Beef and fine Biscuit. Colonel White, having provided an ample provision for his whole detachment, our men fared sumptuously and towards evening we fell back to a swamp 2 miles in our rear for the night (this was on the 28th. June 1778 the day of the Battle of Monmouth). At day break a large decked boat full of men appeared under sail in the Rassau passage, and made a signal for the Bluff, which not being answered, she put about and hastened down the rivers with oars. Captain Murray had notice of this by one of the Loyalists whom he had left there as a decoy and joined Major Prevost at 10 o'clock. At about mid-day when the men from the detachment were cleaning their arms and others washing in the creek in front of the camp the Grenadiers constructing a breastwork,—Major Graham's detachment had joined on the evening before,—while Colonel Brown encamped at a swamp in front about 10 miles to collect his men,—at mid-day a mounted Sergeant came to Major Prevost's tent to announce the Colonel being at hand with his Corps. The Major, by the Sergeant, desired Colonel Brown to come to his tent; presently we heard Whitfield the Drummer beating the Grenadiers' March, and the rangers filing over a bridge where he had a sub-guard. Soon after entering when the Ranger drum beat to arms, and the Rangers many of whom had lost their arms in the cabbage swamp, were seen flying into that in front of our camp, Captain Muller [1] of the Grenadiers and his men running to the Camp for their arms, musket balls whistling over our tents while the enemy with sabres and rifles were shouting, 'down with the Tories.' Captains Smith and Johnston with a few of their Rangers bravely defended the Aligator bridge, till the regulars having got under arms and relieved them, the enemy, disconcerted by their cool and deliberate fire, and General Scriven being wounded, they fled to their horses, leaving one Ensign on the field dead, and a black man—the General's servant—well stored with [2] . . . prisoner. Their loss is not supposed to be great, as they only stood from behind trees two or three fires from the troops. One man of the 2nd. Battalion was killed and one of the 4th. wounded;—Captains Johnston and Smith with five of Brown's rangers wounded. The horses being at pasture could not be collected during the short action, and Ensign Gigilighter of the South Carolina Royalists who wore yellow breeches were taken at the advanced guard and caried off. Major Prevost did not allow any pursuit; the surprise was complete on both sides, for Scriven did not expect to find us there. 8 or 10 Indians made their appearance after the firing was over; they came from St. John's River, and were not allowed to pursue. After sunset Captain Murray's Light Company was advanced to

[1] 3rd Battalion. [2] Illegible in MS.—L. B.

reconnoitre in front; they found some sacks filled with staples in a swamp near the road side which the men carried to Camp. Next morning Major Prevost with Colonel Brown, Captain Murray and a party of South Carolina Dragoons advanced on horseback about ten miles where they found some of the enemy repairing a bridge over a Creek on the upper road. On our approach they left their tools and fled to their horses, and being much better mounted than our men Captain York could not come up with them, but burnt the bridge and carried off the tools. [That] day Major Prevost retired to Six Mile Creek, felling the trees on the road through the swamps, leaving Indians to observe the enemy. General Prevost hearing of the intended movement of the enemy had prudently withdrawn his advanced corps which was more than one-third of his force. 2 frigates guarded Amelia [on (?)] St. Mary's Sound and cruised along the coast making several prizes, while our Rangers and Indians sent in Cattle and Horses, as also provisions. Colonel Brown surprised Fort Barrington on Altamaha and took the garrison of 20 or 30 Georgian continentals.[1] Lieutenant True Drew of the Rangers was killed.

'In the beginning of November 1778 Major Prevost with all the Cavalry, East Florida Rangers, South Carolina Royalists and Mc Girth's men, with the Grenadiers of the 2nd. Battalion, and 70 chosen men of the 3rd, amounting in all to 750 men with a 4½ inch Cohorn mounted on a Congreve Carriage [issued orders as follow]:—

'The Cavalry to proceed to Sapelo to wait for the Infantry, and from thence to proceed jointly to Medway, to attack Sunbury in conjunction with Lieutenant Colonel Fuser, who proceeded along the inland communication with 250 men of the 4th. Battalion, the armed *Flat Thunderer* of 2 24 pounders, and 2 Swivels; while the Privateers *Spitfire* and *Aligator* were to alarm the Seaboard.

'November 24. Colonel Fuser landed at Colonel's Bluff at the mouth of Newport, where he learned that, 2 Privateer's men having deserted and given the alarm, 300 men had been marched to Sunbury. The Colonel mounted the 2 Swivels on a cart, by way of carriage and leaving 60 men to guard the boats, proceeded towards Sunbury with 180 men receiving shots from their look-out men who fled to the woods whenever Ensign Shoedde[2] with the flanking

[1] Continentals: i.e. men of the American regular army.
[2] Ensign C. L. T. Schoedde of the 4th Battalion.

party advanced upon them. There was no firing on our part except by 3 or 4 of Brown's Rangers acting as guides. When we came to the marsh which divides the Island from the main, and is passable at low water, the detachment was ordered to form at open order, there being wood beyond the marsh, and the Medway on the right. Captain Murray was ordered forward to cover the left flank and clear the wood: Captain Wulf with his Grenadiers to support them. No enemy appearing, the D[etachment] proceeded in open order over the marsh, and then in a column of files. At a turn of the road 2 officers mounted fired rifles and narrowly missed Colonel Fuser, so that [as on] approaching the town, the [American] flag was seen displayed on the Citadel, the D[etachment] marched on towards the left among some small trees to the Newport road, when in answer to our Drum, a shout was heard of 'God save King George,' and whooping like Indians moving towards Medway meeting.

'We bivouacked at night on the slope of the rising ground opposite the fort and made fires in our rear which was considerably more elevated, so that when our Drums beat the Retreat they fired several cannon shot at our fires over our heads. This salute being performed, Colonel Fuser and Captain Wulf went close to the Fort to reconnoitre. They found it well provided with heavy guns and men, but no appearance of a gate on that side; that towards the sea was known to have a battery of 18 pounders. Captain Murray was sent with his Light Company to try if he could not get into the town, which he did from the Medway Road, where we met with a lighted house outside the town, where we found our old Nassau acquaintance the Doctor and his two daughters. He said he would not leave his house as all the town had done, because [he] considered himself on parole.

'On entering the town a light appeared at the water-side, which we took for a guard at the wharf. Ensign Schoedde's party advancing in the dark to take it, found it to be a galley at anchor, it being high water, but the night being very dark and our men silent they were not seen and joined us in the Courthouse which Captain Murray had taken post in. Lieutenant Campbell was despatched to the Bivouac and Colonel Fuser came in. As the Court house was not finished Colonel Fuser took possession of a Merchant's house where a puncheon of rum was broached, with other refreshment he distributed among the D[etachment] but no plundering allowed. Although Captain Wulff patrolled the town up to the Citadel without finding a gate, only 2 men were found, all the rest having taken refuge in the Fort ; every now and then they fired great guns at our fires while our men occupied six houses with stores for 18 hours.

'Captain Johnstone of Brown's Rangers pushed on to Medway meeting, and returned in the morning reporting that Major Prevost had pushed on to Ogeechee. Ensign Schoedde was posted 3 miles on the Medway road to look out. Colonel Fuser summcned the Fort allowing an hour ; in two hours Colonel Mc Intosh sent Major Lane with a spirited answer. An American Detachment entered by the Newport road and it being highwater, Colonel Fuser would not suffer us to attempt to storm the Fort, but drew out the Detachment until Ensign Schoedde's party was with-drawn ; the Light Company in front of the line at open order. The D[etachment] then filed off by Medway road [after] a few shot from our Rangers, and the Light Company closing the march and leaving the astonished enemy who durst not disturb us. When we turned to the left and passed the pond behind the town two or three shots were fired at us, [but] we proceeded to our boats without any interruption. There he received a letter by a Ranger from Major Prevost that General Scriven was killed and that he [himself] had retired to Newport with his prizes, negroes, horses, cattle and other effects, that he had broken down the bridges in his rear, and could not join to return to Sunbury. Colonel Fuser's D[etachment] crossed to the S[outh] side of Newport. Captain Murray went up the river with one man in a canoe, landed one mile below the bridge, and found them with Geese, Turkeys, Pigs, Fowls &c : roasting around their fires. Major Prevost read a letter to Captain Murray purporting to have been dropped on the road, which he allowed Captain Murray to read but would not let him copy. His prizes had filed off to the rear. Captain Murray rejoined Colonel Fuser about 1 in the morning and copied the letter from memory ; when compared with the original afterwards, there was only two words different.

'Major Prevost when on St. Mary's North side had a public sale of his booty which amounted to £8000 and halted at St. John's river. Colonel Fuser fell back 1st. to Sapelo next to Frederica ; then to Jekyl, covering the retreat of Major Prevost's division. After all his craft had passed down the Altamaha Colonel Fuser crossed St. Andrew's Sound, and at the Narrows he received orders to wait at the south end of Cumberland for General Prevost and all the troops of the Garrison except 4 Companies of the 3rd. Battalion who were to remain with Major Glazier. General Prevost was also accompanied by 3 Companies of the Jersey Volunteers who had parted company with Colonel Campbell's Division from New York, under L. C. Allen.

'The Army proceeded from the South end of Cumberland Island, but the Quarter Master General Prevost having disposed of the Cattle and other provisions the army was reduced to subsist on oysters, and a very scanty supply of rice from the new made Commissary Fatia. On the 2nd. [February 1779] afternoon a gale of wind in the South between Cumberland and Jekyl drove back all the flotilla to the old port St. Andrew's except Ensign Finlay's canoe of the 3rd Battalion which reached Jekyl, and Major Prevost's, Captain Murray's and Captain Muller's which reached little Cumberland. One boat of the Jersey Volunteers joined Major Prévost from a desert Island between the two Cumberlands, and for three days we were subsisted on an Aligator and some Madeira wine from a Ship that had been wrecked on the coast. Messrs. Birch of Liverpool to whom the vessel had belonged presented the troops with a pipe of Madeira for collecting those that lay scattered along the Beach. When our boats reached Jekyl we were served with horseflesh.

'The General despatched Major Graham with the 3 Companies of the 16th., and Captain Murray's Company with Ensign Schoedde's party to take the rice stores on Broughton Island and then to push on for Sapelo high bluff where they were to wait for the General and in the meantime kill cattle for the army. Ensign Schoedde went up the south bay of Sapelo and made the guard of 7 men prisoners. Captain Murray going to support him, met him with his prisoners, and they all followed the Major by the inland passage. When they had got on about 6 miles, it being moonlight, they heard voices approaching with oars, but soon discovered it to be Major Graham who had mistaken the passage and was obliged to retrograde into Doboy Sound to rectify his mistake. Captain Murray pushed on to Sapelo Indigo work, whence taking 30 men with Baron Breitenbach[1] and Ensign Schoedde and not finding any Gallies at the Salt Depot of the enemy, they proceeded along the avenue that led to the house, keeping in the shade. Ensign Schoedde entered their guard house, and made the guard Prisoners without any noise. Sergeant Dornseif and two men changed clothes with the prisoners, and very cooly mounted the stairs to the look-out at the top of the house and relieved the Sentry. Then Lieutenant David Montaigut the Commandant who with all in the house, the sentinel excepted,

[1] Lieutenant in the 4th Battalion.

were fast asleep, was waked; two young Georgian Officers went out with Ensign Schoedde and Campbell to catch horses and hunt cattle. About 2 P.M. 26 head of cattle were brought in and slaughtered for the army. [Then] Ensign Schoedde set out for Colonel's Bluff, but by mistake ran up Newport River; when Major Prevost arrived near the Bluff with Captain Moncrief the Engineer they were challenged and fired at. They therefore lay upon their oars until the 4th Battalion Light Company came up and landed, when the enemy's look-out galloped off. Captain Murray advanced to the wood where he formed the chain at extended order to cover the landing. In less than an hour the 16th. and 4th. Battalion Light Company were despatched to Sunbury, where our Cavalry were supposed to be investing the place; but none appearing Captain Murray entered, and the advance took post in the ditch of the intrenchment which covered the town. A ranger guide reconnoitring too near the fort was killed, and we took post in the ditch of the entrenchment, opposite the Fort.

'The next morning 23 horses were sent out of the fort when a Sergeant with a few men drove in the escort and the horses were captured. Soon after Mr. Roderic Mackintosh accompanied by his faithful Negro Cyrus, disdaining the counsel of Cyrus, walked under the musketry of the Garrison, setting them at defiance, when they shot him down and disarmed him so quickly that Lieutenant Baron Breitenbach and Sergeant Supman of the 4th. Battalion Light Infantry who with alacrity ran to his rescue could only carry him in wounded in the face. As soon as our men seized him the Americans ceased firing.

'General Prevost came that day with the remainder of the troops, two 8-ineh Howitzers and a Cohorn. The gallies when the tide was high, fired into the town, as did the fort. On the 3rd. day the enemy attempted a small sally which Major Graham drove in; three men of the 16th were wounded, not dangerously. Sergeant Balany Royal Artillery threw some shells at the gallies, which dislodged them, and a shell fell upon a building where the rebel Officers messed, and killed and wounded 9 of them, and shattered about 50 stand of arms; upon which they proposed to capitulate; which being refused and 2 more shells falling into the fort, they hauled their colours down and surrendered at discretion. Captain Moncrief received a contusion in the breast. The gallies made for the Bar: Captain MacDonald[1] mounted guard in the

[1] Captain Donald McDonald of the 3rd Battalion.

Fort with the 3rd. [Battalion] Grenadier Company and next morning Captain Wulff relieved him with those of the 4th. [Battalion]. Captain Macdonald delivered his report to the General and dropped down dead at his feet. The Garrison with their Commander Major Lane embarked for Savannah. They, with prisoners brought in by our mounted rangers might amount to between 300 and 400. Lieutenant Colonel Allen was left at Sunbury with the Jersey Volunteers. Mr. Mackintosh was appointed Captain of the Fort, he lost the use of his eye.

'The Flank Companies except the Grenadiers of the 4th. who escorted the prisoners by the inland passage marched to Savannah by Medway and Ogeechee, and the General was escorted by a party of Dragoons from Colonel Campbell's mounted Infantry; while our Rangers scoured the country in their front. One galley blew up on Sunbury bar; a sloop and 2 galleys were taken by our cruisers and a ship in the harbour.

'At Savannah we learned that our General was promoted to the rank of Major General, and our troops were called the Florida Brigade under Lieutenant Colonel Prevost; and Major Gardner 16th. commanded our three Light Companies, and Major Glazier the Grenadiers; Major Van Braam the Battalion men of the 16th. and 60th. Colonel Fuser being senior to Colonel Campbell was sent to command at Augustine. 2 days after our arrival at Savannah, Colonel Compbell wàs detached with about 1000 men who advanced to Augusta, but hearing that the American General Williamson was advancing with a superior force he unaccountably fell back to Hudson's ferry, and soon after embarked for England, with Commodore Hyde Parker in the *Phoenix* with a rich prize ship loaded with indigo.

'In the meantime the 3 Florida Light Companies under Major Gardner made an excursion to Abercorn and Purisburg by the River Savannah to alarm the enemy and returned next day. The 2nd. day after this our little Corps embarked in 4 Transports and dropped down to Tybee from whence escorted by the *Vigilant* of twenty-four 24-pounders (Captain Christian) and the *Germaine*, (Lieutenant Mowbray) by the Passage inside of Hilton Head Island into Broad River to make a show by way of diversion to favour Colonel Campbell's operation. After passing the strait which leads into Broad Sound, we landed on Hilton Head and pursued Captain Dogherty whose party had fired upon us as we passed the strait, and burned two houses. The vessels proceeding up Broad River anchored opposite the elegant house of General Bull on the Island of Port Royal. Captain Murray was detached with his company up a navigable creek on the South side with orders to burn the plantations whose masters were absent. They landed at a plantation where the master was gone, and with much regret burnt the house of Colonel Hayward who with his sons appeared on horse-back at the edge of the wood, when Captain Murray advanced and called to them to come forward and save the buildings. In answer to this they fired at him, and galloped off. Captain Murray notwithstanding ordered all the furniture to be taken out, and took upon himself to preserve the Overseer's house on account of the Ladies of the Family.

'Lieutenant Baron Breitenbach went to an opposite plantation, whose master, having the gout, the house was saved and nothing taken away. Two armed negroes of Colonel Hayward's came under the bank of the Creek skulking for a shot, but were hemmed in by Sergeant Birnie and two of the men to whom they surrendered. A tierce of indigo was brought off, but no plunder allowed from the house.

'In the evening when the Company joined their ship, a body of Virginians who were posted in General Bull's house were firing rifles and musketry at our ships and grossly abusing the people. Next morning the *Germaine* cannonnaded the house and dislodged the enemy. The troops landing at the same time, hunted them into the wood where Baron Breitenbach formed a chain which kept them at bay. No furniture was left in the house but in the attic storey at the top of the house, where likewise was a Cupulo, a billiard table with all the apparatus, the shelves for a Library along the walls of the room but no books, the lockers contained 120 dozen of liquors; the ale was served to the men and the wines sent on board, after which the buildings were sent on fire and the detachment re-embarked; it being high water. A Sergeant, Corporal and Drummer of Rebel artillery hailed us, and informed [us] that Captain Treville's Company of Artillery had evacuated Port Royal, and that the explosion we had seen the smoke of, was that of the Magazine.

'In the evening a council was held upon Captain Murray's proposal to attack and seize Port Royal. Major Gardner, Captain Murray and Captain Bruère,[1] being in favour, and Captain Christian, Major Graham and Lieutenant Mowbray being against, Major

[1] Of the 3rd Battalion.

Gardner did not think fit to insist, and dismissed the Council to their ships which moved up the River, agreeable to instructions. It was in vain to urge that the General could not foresee that Port Royal, the only harbour where a ship of the line could be received south of the Chesapeake, would with so little trouble be in our power; and Captain Murray's offer to hold the place with his Company was rejected. Captain Murray was to join the General as Aide-de-camp on his return to Savannah, and was entrusted with secret intelligence by which he had reason to believe that several loyalists would declare themselves and assist in defending so safe an asylum as Port Royal, which the *Vigilant* would have secured from all attempts.

'The ships were moved a few miles up the Sound, and anchored. About midnight a negro came in a canoe alongside the *Margaret and Martha* Transport and reported that General Ashe with 1,100 men was at Cupohatchie[1] bridge on his march to Georgia, and that he had left 300 sick at Pocotalalifo[2] bridge from which the negro had made his escape. Captain Murray immediately took the negro with him in his canoe on board the *Vigilant*, reported this intelligence, and proposed with the Flood [tide] to attempt a surprise, or at least to take their hospital; that either of these would derange General Ashe's movement. The Major declined the proposal alleging with truth that Captain Christian thwarted him in his operations. Captain Murray returned quietly on board the *Margaret and Martha*.

'At daybreak of the 3rd February 1779 the signal was made embarking in boats 13 minutes after, for rendevozing [*sic*] alongside the *Vigilant*; and as soon as assembled, the 3 Companies landed at a house where a white flag was flying (for Captain Christian refused the assistance of his 40 marines). The Cohorn[3] was also landed with 2 Artillery men and 6 sailors. A very indifferent horse was procured for Major Gardner. The Corps moved on; Lieutenant Calderwood having the advance and Baron Breitenbach the rear guard. After 2 miles march some shots were exchanged with the American Dragoons, who were driven by Lieutenant Calderwood to the road leading from Roupels ferry to Beaufort, towards which place were seen marching some carriages with an escort indistinctly seen on account of the dust. Lieutenant Calderwood 16th. was sent to reconnoitre them; who soon returned, and reported them to be only a sergeant and twelve escorting

[1] ? Coosawhatchie.   [2] ? Pocotaligo.

[3] A Cochorn—so-called after the inventor—was a small mortar.

baggage. Major Gardner then disposed the Corps into 9 platoons as under :

Major Gardner
(Cohorn)
Lieutenant Calderwood 16th.
Major Graham ,,
Lieutenant Skinner
Lieutenant Plummer (Plumer) 60th.
Captain Bruère ,,
Lieutenant Finlay ,,
Lieutenant Hesselton (Hosleton) ,,
Captain Murray ,,
Lieutenant Baron Breitenbach ,,

The Major then turned off towards Roupelles Ferry leading from the centre and fired the Cohorn at some Dragoons who were observing us on the road, and following them up Captain Bruère with his platoon of marksmen, threw himself into the side of a road where 2 horsemen were advancing, and Doctor Fraser of the American Militia coming up with his negro man were taken with medicines and instruments. Captain Murray knew Doctor Fraser to be a loyalist, and told him so, but he begged to be treated as an enemy, lest a deserter should inform against him. The Major therefore proposed to change horses ; the Doctor being very well mounted, preferred walking.

'About a mile farther on, the Corps halted at a spring, where Dr. Fraser parted that morning with General Moultrie with at least double our numbers and three pieces of cannon, so that what we saw at Rhodes Swamp was his rear, and not a mere Sergeant's party. We proceed[ed] to a causeway over a swamp and came in view of Roupelle's ferry where 2 large Boats full of men were crossing, with tents for about 200 men with the rebel flag flying. As soon as our van appeared out of the wood the boats put back, and they cannonnaded us with two 6-pounders. The Major did not think it proper to bring the rear out of the wood but withdrew by files, when Baron Breitenbach, who had the advance in the retrograde movement, finding some of the enemy's Dragoons demolishing the bridge, charged and dispersed them. They returned on horseback, and the Baron skirmished with them along the road to the entry of Rhodes' swamp, where—on the crest of the Pina Barren beyond the swamp where the trees were felled but not cleared off,—were distinctly seen, the Americans with 3 pieces of cannon: a Company of Artillery, the Virginia Riflemen, the silk stocking Company of Charlestown, all gentlemen, and other Militia; the whole under Generals Moultrie and Bull; besides Captain Barnwall's Dragoons. We think they could not be less than 400 or at least 350. It was evident that they did not wish to bring on an action by their position to defend the only pass on the road, and leaving the way to our ships open to us; but the Major resolving to force the causeway and attack this force on ground he was unacquainted with,—(regardless of the remonstrances of Major Graham and Captain Murray) with a magnanimity worthy of the hero of Cervantes—though his whole force did not [number] 150 or 160, officers and Gunners included,—gallops along the causeway with a white handkerchief at the point of his drawn sword and orders them to lay down their arms and send him an officer. Lieutenant Kinloch came forward, and said they had too much British blood in their veins to yield their post without dispute, and that they knew they were 2 or 3 times superior in number; at the same time refusing to name his Commander. The Major desired Lieutenant Kinloch to cheer with his corps as a sign of refusal. This was complied with, and instantly the Cohorn discharged its contents which killed the American Captain of Artillery and [was] answered by their 3 pieces which killed Lieutenants Calderwood and Finlay, and also the Major's horse under him.

'In the meantime Captain Murray formed left wing under a heavy fire of grape and small arms. Major Graham found more difficulty on the Right from the nature of the ground. Captain Murray received a grape shot in his right haunch as he was dressing the front rank to receive the enemy who was advancing in a false direction. Lieutenant Breitenbach came to report that he had two of his men wounded in attempting to pass the swamp to the left. Captain Murray ordered him to retire his platoon and pass the causeway and form on the left. Soon after Captain Bruère, who with his marksmen had silenced the enemy's cannon, came running to the rear where the doctors were left in a log house, declaring he was wounded through the ribs; but Sergeant Fitzgerald maintained the action till he was shot through the body. Our Cohorn was left at Rhode's house, the sailor who carried the match running away after the first round. Major Gardner narrowly escaped being taken by Captain Barnwell's Dragoons, by the American Artillery Sergeant &c. telling them they would take care of the Major whom they led into the swamp in rear of the right with

Doctor Clarke.[1] Major Gardner then sent off Major Graham with his platoon to escort the Cohorn to the ships. Major Graham had received two buckshot in his thigh. Major Gardner then ordered the retreat to be beat, and sent Corporal Craig 16th. to Captain Murray with orders to retire. Captain Murray ordered his men to advance, and desired the Corporal to say that the left wing was driving the enemy and that a retreat through a defile in the rear was impracticable. Corporal Craig then said that Major Graham was killed and Major Gardner a prisoner. Drummer Hynes 16th. beat the advance and Craig was despatched to order Lieutenant Skinner [to] extend to the right, skirmishing by platoons from left to right; while the left wing did the same from right to left leaving the center open; and when the flank Platoons had gained [the] enemy's flank they were to charge in upon them. This was executed promptly on the left. The enemy withdrew 2 of the guns and moved the other to their right to support the riflemen who were so briskly pushed that the cannon was never fired, but speedily withdrawn. After the riflemen had given way a single rifleman remained behind and shot Captain Murray through the left arm—as he was waving it for the Baron to charge their flank while he was doing the same in front—and dropped out between the shoulder blade. This occasioned a pause by which the enemy having horses and train, carried off their gun. When the left wing charged the riflemen threw away their arms though the charge was made at open order. The rebel ammunition being expended they hastened to the road and retired in confusion, pressed by both wings who did not fire, [they] having little ammunition left; and Lieutenant Skinner, after following up their left wing to the road, drew up the men facing the enemy. At this moment Captain Murray recovered from fainting, and assisted [by] some wounded men, joined the little Corps now reduced to 63 rank and file: Captain Murray and Lieutenant Plummer both wounded; Lieutenants Skinner and Baron Breitenbach, with 6 or 7 sergeants and Drummers fit for duty; only 93 cartridges remaining.

'It was thought advisable to collect the ammunition from the killed and wounded—which produced about 300 cartridges from the dead on both sides—to compliment the retreating enemy with 3 cheers which they did not return; and after collecting our wounded, the sun being now set, we moved with advanced arms

[1] The meaning of this is explained by the sequel. The Major seems to have been with some Americans who had just deserted to us. Barnwell's Dragoons naturally thought he had been captured by them.—L. B.

from about 600 yards in front of where the action began. Lieutenant Skinner with the marksmen drove away some Dragoons who were observing us. At Rhodes' house were found Dr. Fraser, with Lieutenant Hesselton shot through the groyne, Sergeant Dornseif, his leg fractured by a cannon ball, Sergeant Fitzgerald, and several others.

'Captain Murray placed Lieutenant Skinner in front of the line of march with the 16th.; the wounded, who could walk, in the centre, escorted on the road sides by the 3rd Battalion. Those of the 4th. Battalion to bring up the rear with the ammunition. In case of an attack in front Lieutenant Skinner was to charge, the night being dark and it being supposed that those repulsed from Roupels ferry must have heard the noise of our action. Doctor Fraser was released on condition of attending our wounded. Close to us appeared some horsemen who retired as we advanced, and upon ascending the bank from the swamp, a voice from the right cried 'Fire my lads' upon which the 16th. and 3rd. [Battalion] Company fired at the Dragoons who galloped off. A white horse without a rider came into the rear, and Captain Murray seizing his bridle was assisted to mount him. Captain Bruère and 15 more who had been prisoners to Captain Barnwell were retaken; such as could walk were placed in the center with the others. Captain Bruère mounted behind Captain Murray, and those who could not walk were lodged under Doctor Fraser's care in the log house. Captain Murray was the only man who knew the road in the dark. As they proceeded to the ships, every now and then a man or two emerged from the Swamp on our left, and at last Major Gardner and Doctor Clarke who had got so far along the edge of the Swamp guided by the rebel artillery deserters. The Major imagined that we were all taken. He had received a shot through his lapell; joined with his drawn sword and silk stockings, perhaps Doctor Fraser's Pistols and Spurs. He was lavish in praise of the detachment, and active in assisting the wounded. Within half a mile of the Ships the 40 marines were lying down in a hollow. Captain Christian had landed them (upon Major Graham's report of the Corps being defeated) to receive stragglers and keep the enemy off. Lieutenant Mowbray acted as their guide. The wounded were conveyed on board the *Vigilant* where they were dressed, and Corporal West of the 4th.[Battalion], who distinguished himself in the action had his left arm amputated, and composed a song next morning. Antoine Metro of the 4th. [Battalion], a very brave soldier, was shot through the heart as he drew his trigger, skirmishing advancing; and fell touching his Captain. Sergeant Dornseif of the 4th. [Battalion], lost a leg. Sergeant Fitzgerald 3rd. Battalion recovered and was appointed Quarter Master afterwards.[1]

'Next morning General Moultrie sent his Aid de Camp, Captain Day to request bark and other medicines for our wounded and for his, which were much more numerous; that they had buried Lieutenants Calderwood and Findlay, and the rest of our killed with the honours of war, at the same time with their own officers and men, and that Lieutenant Hesselton with our other wounded were under the care of Doctor Fraser. Major Gardner and Captain Christian also wrote to Head Quarters, but their letters never were published, which is the reason the writer has entered into so long a detail of this expedition which had it been conducted by Colonel Fuser or Marc Prevost would probably have produced more important results. General Moultrie's, the only account made public, is as candid as can be expected from a republican. He acknowledges that he could not prevent his men from giving way when their ammunition was expended, but says that we beat the retreat first; which is true as above related, but both parties kept the field for an hour after this, when General Moultrie left us Masters of the ground. That can be attested by Major General Skinner, Baron Breitenbach and Lieutenant Plummer if living, the only officers besides Captain Murray who kept the ground from which the enemy retreated. Yet the Reverend Doctor Ramsay with republican assurance, in his history of South Carolina, calls this a remarkable defeat of the British Light Infantry, though General Moultrie honestly acknowledges his troops retired to Beaufort, leaving the British masters of the ground.

'February 5th. Major Gardner made an excursion among the settlements, and carryed off between 3 and 400 negroes, who had fled from their Masters, with plate and other valuable effects. The negroes were taken possession of by the Quarter Master General and Commissary's Department.

'After this expedition the Light Companies of the Florida Brigade were joined to the 2 of the 71st and 4 Provincial Companies from New York, and placed under Lieutenant Colonel Maitland of the 71st.

'The next remarkable events where the 60th. was employed in 1779 were the surprise of General Ashe at Briar Creek by Lieutenant

[1] William Fitzgerald was Quartermaster to the 1st Battalion, 1784–1786.—L. B.

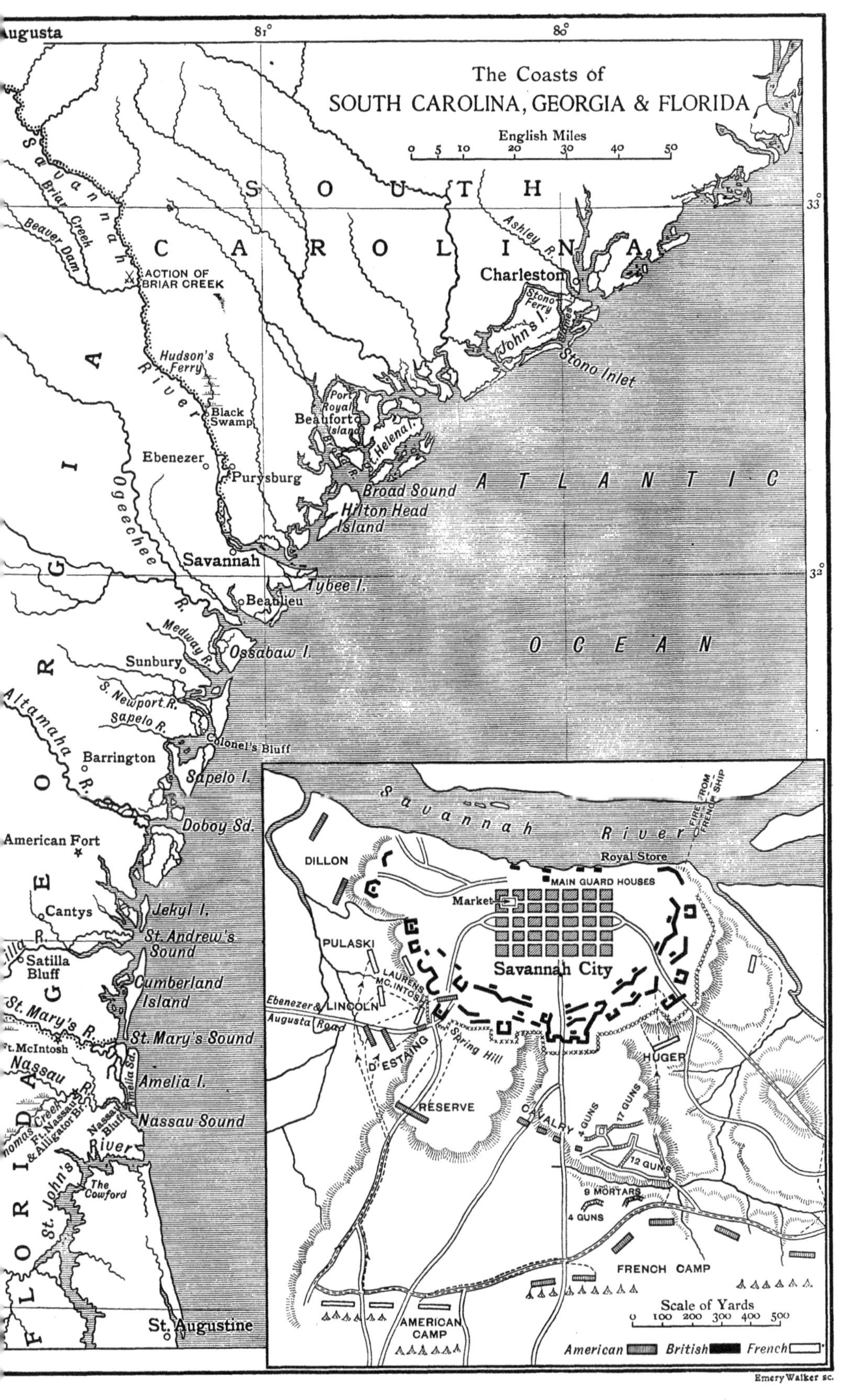
The Coasts of
SOUTH CAROLINA, GEORGIA & FLORIDA
English Miles
0 5 10 20 30 40 50
Augusta
81°
80°
33°
32°
SOUTH CAROLINA
GEORGIA
FLORIDA
ATLANTIC OCEAN
Savannah
Briar Creek
Beaver Dam
ACTION OF BRIAR CREEK
Hudson's Ferry
River
Black Swamp
Ebenezer
Purysburg
Ogeechee R.
Savannah
Tybee I.
Beaulieu
Medway R.
Ossabaw I.
Sunbury
S. Newport R.
Sapelo R.
Altamaha R.
Colonel's Bluff
Barrington
Sapelo I.
Doboy Sd.
American Fort
Cantys
Jekyl I.
St. Andrew's Sound
Satilla Bluff
Cumberland Island
St. Mary's R.
St. Mary's Sound
Ft. McIntosh
Nassau R.
Amelia Sd.
Amelia I.
Nassau Sound
Ft. Nassau & Alligator Br.
Nassau Bluff
River
St. John's
The Cowford
St. Augustine
Ashley R.
Charleston
Stono Ferry
John's I.
Stono Inlet
Port Royal Island
Beaufort
St. Helena I.
Broad R.
Broad Sound
Hilton Head Island
Savannah River
FIRE FROM FRENCH SHIP
Royal Store
DILLON
MAIN GUARD HOUSES
Market
PULASKI
Savannah City
LAURENS
MC.INTOSH
Ebenezer & LINCOLN
Augusta Road
Spring Hill
D'ESTAING
HUGER
RESERVE
CAVALRY
4 GUNS
17 GUNS
12 GUNS
9 MORTARS
4 GUNS
FRENCH CAMP
AMERICAN CAMP
Scale of Yards
0 100 200 300 400 500
American
British
French
Emery Walker sc.

Colonel Mark Prevost, with 3 Grenadier Companies of the 60th, a battalion of the 71st. and some detachments of the Florida Brigade; in all about 900 men, including Sir James Baird's Company, Tawse's Dragoons. General Elbert and 26 officers were taken.'

[*At this point the MS. ends abruptly.*—L. B.]

APPENDIX III

NOTES ON OFFICERS CONNECTED WITH THE REGIMENT DURING THE SEVEN YEARS' WAR

JOHN CAMPBELL, 4TH EARL OF LOUDOUN

1705–1782

JOHN CAMPBELL, 4th Earl of Loudoun, the only son of Hugh, 3rd Earl, and Lady Margaret Dalrymple, only daughter of the 1st Earl of Stair, was born on May 5, 1705.

He succeeded his father in 1731, and from 1734 till his death was a representative peer of Scotland.

Loudoun had entered the army in 1727; was appointed Governor of Stirling Castle, April 1741; and in July 1743, A.D.C. to the King. On the occasion of the Jacobites rising in 1745, he raised a regiment of Highlanders of which he was appointed colonel. The Earl of Loudoun took an active, if not very distinguished, part in the suppression of the rebellion, acting as Adj.-General to Sir John Cope at the battle of Preston, in which engagement nearly the whole of his regiment was killed. After this he proceeded by sea north, and on his arrival at Inverness he raised a second regiment of Highlanders, 2000 strong, with which he relieved Fort Augustus, then besieged by the Frasers.

Loudoun's career as Commander-in-Chief in North America has already been noticed.

In 1762 he was appointed second in command, under Lord Tyrawley, of the British troops sent to Portugal. Loudoun died, unmarried, at Loudoun Castle on April 27, 1782.

FIELD-MARSHAL LORD AMHERST

1717–1797

Jeffery Amherst entered the Foot Guards as Ensign in 1731. He was subsequently present at the battles of Roucoux, Dettingen,

Fontenoy, Lauffeld and Hastenbeck. In 1758, he was promoted to the rank of Major-General, and in the autumn succeeded General Abercromby as Commander-in-Chief in North America. After his return to England in 1763 Amherst gave offence at Court and remained for some time unemployed. In 1772 he was, however, appointed General Commanding-in-Chief, and after a temporary resignation, re-appointed in 1783. In 1776 he had been raised to the Peerage. In 1793 Lord Amherst was given a commission as Commander-in-Chief, and on resigning his office two years later received the rank of Field Marshal.

Lieut.-General John Stanwix

1690–1766

This officer, born in 1690, entered the army at the age of 16 years in the midst of Marlborough's Campaigns, but it was not until 1745 that he gained field rank by promotion to the rank of Major in the Marine regiment. In the same year he was appointed Lieutenant-Colonel in Lord Granby's Regiment, which he seems to have commanded during the 'Forty-five.' Four years later he became Equerry to the Prince of Wales, and in 1754 received the appointment of Deputy Quartermaster-General. In 1756 he proceeded to America, being posted to the Royal American Regiment as Colonel Commandant of the 1st Battalion. In 1758 he was made Brigadier-General, and in the following year promoted to the rank of Major-General while entrusted with the defence of the Western border of Pennsylvania, and allotted the task of subduing the French posts from Pittsburg to Lake Erie. Next year he returned to England, where he entered Parliament as member for Appleby, and became in succession Lieutenant-General and Governor of the Isle of Wight. In October 1766, while on the Staff in Ireland, having completed a tour of inspection, he was crossing from Dublin to Holyhead when his ship was lost, apparently with all on board.

General William Haviland

1718–1784

William Haviland received commission as Ensign in 1739 and would seem to have served at Carthegena and Porto Bello. During the 'Forty-five' he served as A.D.C. to General Blakeney in the defence of Stirling Castle. In 1750 he became Major and in 1752

Lieutenant-Colonel. In 1757–58 he was Commandant at Fort Edward and in the latter year he was present at Ticonderoga. In 1762 he was Second-in-Command to Monckton at the capture of Martinique and commanded a Brigade at the siege of Havana and in the same year was promoted to the rank of Major-General. In 1779 he was in command of the Western District in England. In 1783 he became General and died the following year.

Colonel Charles Lawrence

(?)–1760

Charles Lawrence was appointed Ensign, 1727 ; Captain, 1742 ; Lieutenant-Colonel, 1754 ; Governor of Acadia, 1756. In Loudoun's campaign, the following year, he commanded the reserve with the appointment of Major-General ; he died from the effects of a chill contracted at a ball at Halifax in 1760.

General the Hon. James Murray

1719–1794

James Murray, a son of the fourth Lord Elibank, served at Carthegena and in Cuba and was present at the defence of Ostend in 1745. In 1757 he commanded the 15th Regiment at the expedition of Rockfort. After the capitulation of Montreal, Murray was appointed first Governor-General of the Province of Quebec ; and although not entirely popular with his own countrymen, he held the scales of justice impartially between English and French and thereby won the goodwill and respect of the latter.

On returning to England in 1767 he was appointed to the Staff in Ireland. In 1772 he was promoted to the rank of Lieutenant-General and two years later became Governor of Minorca. In August 1781, with a garrison of 2000 men—a small proportion of whom were regular soldiers and they old and worn out—he sustained a siege from a combined French and Spanish force under the Duc de Crillon. Scurvy broke out in the town ; men actually died on guard, and when the garrison had been reduced to 600 men Murray surrendered. On his return to England he was tried by court martial at the instance of the Lieutenant-Governor, Sir William Draper, and although honourably acquitted was afterwards mulcted in damages to the extent of £5000 as the result of a civil action brought against him for an episode in his career at Minorca. The sum was happily paid by the Government. In 1783 Murray was promoted to the rank of General and died in 1794. He was buried in Westminster Abbey.

LIEUT.-GENERAL THE HON. ROBERT MONCKTON

1726–1782

Robert Monckton, the second son of the first Lord Galway, received a commission in the 3rd Guards at the age of fifteen; and during the year embarked with his regiment for Flanders. He fought at Dettingen and Fontenoy. In 1753 he went to Nova Scotia and in the following year became Lieutenant-Governor of Annapolis. In 1763 Monckton returned to England and saw no further active service.

MAJOR-GENERAL JOHN BRADSTREET

(?)–1774

John Bradstreet was by birth apparently a colonial. In 1746 Bradstreet was Lieutenant-Governor of St. John's, Newfoundland. Ten years later he was entrusted by General Shirley with the charge of forty companies of boatmen—each company containing fifty men—and directed to keep open the communications between Schenectady and Fort Oswego on Lake Ontario. He repulsed the French who attempted an ambush, and his vigour and audacity were conspicuous in his doings—successfully carried out—against Fort Frontenac, immediately after the repulse at Ticonderoga. In 1772 Bradstreet was promoted to the rank of Major-General and died at New York in 1774, his name being retained to the day of his death on the list of regimental captains.

MAJOR-GENERAL AUGUSTIN PREVOST

(*Born* August 22, 1723; *died* May 6, 1766)

Augustin Prevost was born at Geneva upon August 22, 1723, of a French family long settled in Switzerland. He was educated as a soldier, and there is a story which seems apocryphal that he entered the British Service as Cornet in 'the Earl of Albemarle's Regiment of Horse Guards,' a corps which does not appear to find a place in the Army List! Anyhow he served against the French and was wounded at the Battle of Fontenoy. Upon January 9,

Y 2

1756, when in the Dutch service, he was transferred as Major to the newly-raised 60th Royal Americans. His services and wound, obtained at the siege of Quebec, are noticed in the text. On recovery, in 1760, he rejoined the 3rd Battalion in Canada, and having been promoted to the rank of Lieutenant-Colonel on March 20, 1761, took the battalion to the West Indies where, under his command, it played a distinguished part at the capture of Martinique and Havana in the following year. In 1763 Colonel Prevost took the cadre of the 3rd Battalion to England where it was disbanded.

Upon the death of Colonel (Brigadier-General) Henri Bouquet, in 1765, Augustin Prevost was appointed to command the 1st Battalion, first in North America and subsequently in the West Indies.

In 1775 Prevost was sent to Europe to assist in raising once more the 3rd and 4th Battalions, and was appointed Colonel-Commandant of the latter with the rank of Colonel. On his return he took command of the troops in East Florida.

In 1778 Colonel Augustin Prevost, as mentioned in the text, was directed to command the expeditionary force organised for operations in Georgia and Carolina against the insurgent American Forces. On February 19, 1779, he was promoted to the rank of Major-General, and with the able assistance of his younger brother, Lieut.-Colonel Marc Prevost, conducted with much vigour and ability the ensuing operations already described.

Although gifted with a strong constitution and with an indomitable spirit, the long series of campaigns had begun to tell upon Augustin Prevost's health and vigour. Great heat and the incessant worry of this most trying campaign, following upon the unhealthy climate of the West Indies, were now exacting the inevitable penalty. 'I begin to feel the effect of age,' he writes in a spirit of temporary despondency, 'and find that this campaign necessitates the greater physical powers of a younger man.'

Yet his greatest claims to distinction were still to come. The defence of Savannah has been already described, but his correspondence shows a fund of determination and self-reliance on the part of Prevost. It shows, moreover, that a good and soldierlike feeling existed between the French and English. The French prisoners left at Savannah upon the departure of the French Fleet received the kindest and most generous treatment on the part of Prevost. 'La nation française,' writes Le Chevalier du Romain on December 8 upon being exchanged, 'vous doit leurs remerciments pour tout le bon traitement et toutes les bontés qu'ont reçu vos prisonniers,'—'kindness,' adds the Frenchman, 'which all know who have had the honor of fighting you as an honorable foe.'

General Augustin Prevost received much praise for the success of his operations, which proved a happy phase of a campaign otherwise disastrous to British arms.

The following document is a curious testimony of the feelings of the Provincial troops serving in his command. It has been preserved with care among his papers, and may be taken as indication of a spirit of confidence and mutual esteem between the rough colonial forces and the officers of the Regular Army under Prevost, which did not always exist at this period.

The Address of the Officers of His Majesty's Troops in Savannah, in the Province of Georgia, to Major-General Augustin Prevost, &c., &c.

'We beg leave, Sir, to express the high sense we have of the polite, disinterested and impartial behaviour manifested by you to all His Majesty's Troops, since we have had the honour of serving under your command.

'It is with regret we now take leave of you, and do assure you, Sir, we shall always recollect with pleasure and gratitude that in the many trying and distressing situations we have been exposed to, from the nature of the service, and an inhospitable climate, nothing has been omitted by you that could contribute to our ease and comfort, or that of the men under our command. We most sincerely wish you and your family a safe and agreeable passage, and flatter ourselves that such marks of royal approbation will be conferred on you, by our most gracious Sovereign, as your zeal for and success in his service so justly merit.'[1]

Savannah, 27th May 1780.

Thos. Bowden, Maj. Commandg. 2nd Batt. Delancy's for self and officers.

T. Brown Lt.-Col.-Commt. King's Rangers for self and officers.

J. H. Cruger Lt.-Col. 1st Batt. Gen. Delancy's Brig. for self and officers.

Isaac Allen Lt.-Col. Commandt. 3 Batt. N.Y.V. for self and officers.

James Wright Junr. Major Commandt. Geo. Loy.

In 1780 after having served for twenty-two years in North America and the West Indies, General Augustin Prevost returned

[1] They 'flattered themselves' in vain.

to England and purchased the estate of Greenhill Grove near Barnet, where he died on May 6, 1786, aged sixty-three. He had married in 1765, at Lausanne, a Swiss lady, the daughter of a Monsieur Grand, and left three sons and two daughters, the eldest of whom, after serving as an officer in his old regiment, the 60th Royal Americans, became subsequently Governor-General of Canada, and was created a baronet.

NOTE.—The present Sir Charles Prevost has an oil painting of Major-General Augustin Prevost—showing on the forehead the circular scar where the skull was trepanned after his wound.

LIEUT.-GENERAL JACQUES PREVOST

1725–1778

Jacques Prevost, younger brother to Augustin, although senior to him both in the Dutch and British service, began his military career in the Sardinian army. He afterwards joined that of Holland and became a Major in 1749. His career in America on his transference to the Royal American Regiment as Colonel-Commandant of the 4th Battalion was short. He returned to England in 1760, but was subsequently appointed Governor of the island of Antigua.' Prevost was promoted to the rank of Lieutenant-General in 1772 and died at Breda in 1778.

COLONEL JEAN MARC (MARCUS) PREVOST

(*Born*, 1736 ; *died about*, 1780)

Jean Marc Prevost was the youngest brother of Lieut.-General Jacques Prevost and Major-General Augustin Prevost. Like his brothers he entered the Dutch service and with them joined the Royal American Regiment as Captain on January 17, 1756. When the regiment was reduced from four battalions to two Marc Prevost was placed on half pay, but was promoted to the rank of Major in July 23, 1772, and brought again into the regiment upon its augmentations in 1775.

Upon August 29, 1777, Marc Prevost was promoted Lieutenant-Colonel in the army while still serving with the 1st Battalion as Major. As a Lieutenant-Colonel in 1778–79 he greatly distinguished himself under the command of his brother Augustin during the successful operations in Georgia and Carolina.

Marc Prevost's brilliant conduct at Briar Creek gives evidence of dash and capacity; and, during the campaign, at the head of a small improvised mounted corps called Rifle Dragoons—the prototype of the modern mounted infantry—mainly composed of men from his own regiment, the 60th, he showed military qualities and energetic leadership of a very high order. He was also present with his brother at the Siege of Savannah.

About the year 1780 Marc Prevost rejoined the 1st Battalion in Jamaica where he died from the effect of wounds.

He had married Theodosia, daughter of Theodosius Bartow, a distinguished citizen of Shrewsbury, New Jersey, and left a family. His descendants now live at Lima, Peru.

Lieut.-General James Robertson

(?)-1788

James Robertson entered the army as a private soldier in 1720. Gaining a commission he served as an Ensign at Carthegena. On the raising of the Royal American Regiment he was given a Majority therein and in 1758 Brevet rank of Lieutenant-Colonel. In 1776 he commanded a brigade at the engagement at Long Island in the War of Independence. In 1779 he was Governor of New York, two years later was appointed Commander-in-Chief in Virginia. On arrival, however, he found that the appointment had been given from him to Lord Cornwallis and accordingly returned to New York. In 1782 he was promoted to the rank of Lieut.-General, and died in 1788.

Major-General Joseph Frederick Wallet Des Barres

1722–1824

Joseph Frederick Wallet Des Barres was in many respects a remarkable man, and not least so from the fact that his name is to be found for no less than forty-seven years in the list of our Regiment, of which he had been an original member, and which he only quitted at the age of 81, being still a regimental captain—on promotion to the rank of Major-General. Born in 1722, the descendant of a Huguenot who had come to England after the revocation of the Edict of Nantes, he was sent to the Royal Military Academy at Woolwich. Wishing to see active service he embarked in March 1756, with the rank of Lieutenant in the Royal American Regiment to which he was posted as one of the original twenty Engineer Officers attached thereto. Being employed on recruiting duty he collected over 300 men in Pennsylvania and Maryland, and was ordered to form them into a corps of field artillery which he commanded pending the arrival of a battalion of Royal Artillery from England. In 1757 Des Barres commanded a force against the Indians who had raided the neighbourhood of Schenectady and other frontier towns. He surprised and captured the chiefs, but by his tact and ability converted them to friendship, and enrolled them in a corps which performed useful service as auxiliaries until the close of the war. In 1758 he accompanied Amherst to Louisbourg, where he at once performed an important service by landing in spite of violent surf, and capturing an entrenchment. The effect of this was greatly to facilitate the disembarkation of the army. During the siege he acted as an engineer, and afterwards constructed, from the plans and papers found in the fortress, a chart of the St. Lawrence, which proved invaluable to the Fleet at the siege of Quebec in the following year. During that siege he was one of Wolfe's aide-de-camps, and it is stated by Morgan[1] that he was making a report when the General fell into his arms mortally wounded. After the battle of Ste. Foy he was again employed as an engineer and was successful in repairing the fortifications of Quebec which had been almost dismantled. As an engineer he was no less successful in directing the operations against Fort Cartier and other strongholds. After the war he constructed an admirable survey of the coast of Nova Scotia. When the War of Independence broke out the want of charts of the American coast was greatly felt. Des Barres was consequently employed to adapt the surveys of Hollandt and de Brahm (both of them officers in the Royal Americans) for use of the Fleet. He examined and charted the entire coast of the Atlantic, from Florida to Labrador, and his charts are still used by the Admiralty. He published these in 1777 under the title 'Atlantic Neptune,' a copy of which is to be seen in the Royal United Service Institute at this day.

Des Barres seems to have had an unfortunate propensity for quarrelling with everyone he met. Stories of his pugnacious character are still told in Nova Scotia. He was employed as architect in the erection of the church of St. George in Halifax, and directed to construct a square building. Des Barres, however, preferred a round one, and carried his point regardless of all opposition and remonstrance. In 1784 he was made Governor and Commander-in-Chief of the island of Cape Breton. If the story

[1] *Biographies of Celebrated Canadians.*

told of him by Major Murray (p. 295) is correct, it is hardly matter for surprise that he lost his appointment. Being no doubt a poor man, and unable to purchase steps of rank, Des Barres' army promotion had been extremely slow: at the age of 53 he was still a subaltern. At length, after almost twenty years of service, he was promoted to the rank of Captain, and thenceforward rose by successive steps of Army rank until in 1803 he was taken out of the Regimental List, and made a Major-General. His real connection with the Regiment, had however long ceased, for he never did duty with it after the conquest of Canada.

In 1804, at the age of 82, he was appointed Lieutenant-Governor of Prince Edward's Island. At ninety-five Des Barres found himself for the first time with a little leisure. He was still full of life and energy, and made up his mind to spend two years in England, and the three subsequent ones in making a tour of Europe. Whether he executed his idea or was cut off before he could carry it out does not appear. He died at Halifax, Nova Scotia, at the age of 102, on October 24, 1824, having lived just long enough to see the name of his regiment changed from 'The Royal Americans' to 'The Duke of York's Own Rifle Corps.'

Samuel James Holland (or Hollandt)

(?)-1803

Samuel James Holland was one of the Engineer Officers appointed to the Royal American Regiment at the time of its being raised. He effected a large and valuable series of surveys, and at a later date was appointed Surveyor-General in Canada. Holland died in 1803.

APPENDIX IV

COLONEL BOUQUET'S INSTRUCTIONS

THE following pages give the substance of Henry Bouquet's ideas for the conduct of war against the Indians, and certain hints likely to be useful against other savage foes.

PREPARATIONS FOR AN EXPEDITION IN THE WOODS AGAINST SAVAGES

It is not practicable to employ large bodies of troops against Indians; the convoys necessary to support them would be too cumbersome, and could neither be moved with ease nor protected. Several small expeditions are preferable to one that is unwieldy and consequently a force that is intended to act on the defensive should not exceed the following numbers:—2 battalions of Infantry—900; 1 battalion of Hunters—500; 2 troops of Light Horse—100; 1 company of Artificers—20; Drivers and other necessary followers—280: total 1800. The daily ration of a soldier in the woods should consist of a pound and a half of meat and one pound of flour, with a gill of salt per week. Eighteen hundred will therefore require in 6 months 327,600 lbs. of flour, allowance of 25% for accidents 81,900, total 409,500. Including the same allowance for accidents, the meat required would be 614,400 lbs. which would be provided by 2048 oxen at 300 lbs. each accompanying the column; the amount of salt required would be 182 bushels. Of these amounts, however, only one half would suffice for a march from the furthest magazine into the heart of the enemy's country. As regards transport therefore, 1365 horses each loaded with 150 lbs. of flour would suffice; 46 would carry the salt; 50 the ammunition, tents and tools respectively, 20 would be enough for the hospital, and 150 for the Officers' baggage and Staff. This gives a total of 1731 horses, an exorbitant number necessitating great expense. It would be preferable if the country permit of it to use carts each drawn by 4 oxen and carrying 1300 lbs. of flour. One hundred and sixty carts would thus be required drawn by 640 oxen, leaving with the army a reserve of 384 oxen. Transport by cart is less expeditious than by horses, and requires a better road, bridging, &c.; its advantages are the saving of expense and a daily reduction of the convoy by the slaughter of the oxen in proportion to the flour. There are few rivers in North America which cannot be forded by these carts in summer time, and by use of wheeled transport the number of horses required for baggage, ammunition, &c. would be reduced by between 3 and 4 hundred.

Encampment

The camp (figure 1.) forms a parallelogram of 1000 feet by 600: 800 of the regular troops encamp in the four sides, the Light Horse encamp within the parallelogram; the reserve is in the centre; provisions, ammunition, tools, stores and cattle are placed between the two troops of Light Horse and the reserve; the Hunters encamp diagonally at the four angles, being covered by redoubts formed with kegs and bags of flour or fascines. Besides these four redoubts, another is placed to the front, one to the rear, and two before each of long faces of the camp, making in all ten advance guards, 22 men each and 7 sentries, covered if possible by the breastworks of fascines or provisions. Before the army lay down their arms the ground is to be reconnoitred and the guards posted who will immediately open a communication from one to the other to relieve the sentries and facilitate the passage of rounds.

The sentries upon the ammunition, provisions, headquarters and all others in the inside of the camp are furnished from the reserve. The Officers, except the Staff and Commanders of Corps encamp on the line with their men.

The fires are made between the guards and camp, and put out in case of an attack in the night.

Line of March

Part of the Hunters in three divisions, detached in small parties to their front and to their right and left to search the woods and to discover the enemy.

Artificers and axemen to cut a road for the convoy, and two paths on the right and left for the troops.

One hundred and fifty men of the regular troops in two files who are to form the front of the square; these march in the centre road.

Two hundred and fifty regulars in one file by the right hand path, and 250 by the left hand path are to form the long faces.

These are followed by 150 regulars in two files who are to form the rear of the square.

The reserve composed of 100 regulars in two files.

The rest of the Hunters in two files.

The Light Horse.

The Rear guard composed of Hunters follows the convoy at some distance, and closes the march. The scouting parties who flank the line are drawn from the Hunters and Light Horse and posted as in the plan. Some orderly Light Horsemen attend the General and Field Officers who command the grand Division to carry their orders. Two guards of Light Horse take charge of the cattle.

The convoy proceeds in the following order:—

The tools and ammunition following the front column.

The baggage.

The cattle.

The provisions.

The whole divided into brigades, and the horses two abreast.

Defiles

In case of a defile the whole halts until the ground is reconnoitred and the Hunters have taken possession of the heights. The centre column then enters into the defile followed by the right face; after them the convoy; then the left and rear faces with the reserve, the Light Horse and the rear-guard.

The whole to form again as soon as the ground permits.

Dispositions to Receive the Enemy

The whole halt to form the square or parallelogram, which is done thus:—The two first men of the centre column stand fast at 2 yards distance; the two men following them step forward and post themselves at 2 yards on the right and left. The others come to the front in the same manner till the two files have formed the rank which is the front of the square.

The rear face is formed by the two file-leaders turning to the centre road where, having placed themselves at 2 yards distance they face outwards and are followed by their files, each man posting himself at their right or left, and facing towards the enemy the moment he comes to his post.

As soon as the front and rear are extended and formed, the two following faces which have in the meantime faced outwards join now the extremities of the two fronts and close the square. This evolution must be performed with celerity.

To Reduce the Square

The right and left of the front faces to centre where the two centre men stand fast. Upon the word 'March' these step forward and are replaced by the two next who follow them, and so on, by such means that the front becomes again a column, the rear goes to the rightabout, and each of the two centre men leads again to the side path followed by the rest.

While the troops form, the Light Horse in each division of the convoy take the ground assigned to them within the square as if they were to encamp, and the horses being unloaded two parallel lines will be formed with bags of provisions to cover the wounded and the men unfit for action. The Hunters take post on the most advantageous ground on the outside, and skirmish with the enemy till the square is formed, when upon receiving their orders they retire within the square.

The small parties of rangers who have flanked the lines of march remain on the outside to keep off the enemy and observe their motions.

When the firing begins the troops will have orders to fall on their knees to be less exposed till it is thought proper to attack.

The four faces formed by the regular troops are divided into platoons chequered, one half, composed of the best and most active soldiers, is with the first firing, and the other the second firing. The eight platoons at the angles are of the second firing in order to preserve the form of the square during the attack.

It is evident that by this disposition the convoy is still covered, and the light troops destined for the charge remain concealed; and as all unexpected events during an engagement are apt to strike terror and create confusion among the enemy, it is natural to expect that the savages will be greatly disconcerted at the sudden and unforeseen eruption that will soon appear upon them from the inside of the square, and that being vigourously attacked in front and rear at the same time, they will neither be able to resist nor once broke, have time to rally so as to make another stand. This may be effected in the following way.

General Attack

The regulars stand fast.

The Hunters sally out in four columns through intervals of the front and rear of the square, followed by the Light Horse with their bloodhounds.

The intervals of the two columns who attack in the front, and of those who attack in the rear, will be closed by little parties of Rangers posted at the angles of the square, each attack forming in that manner three sides of a parallelogram. In that order they run to the enemy, and having forced their way through their circle follow upon their flanks by wheeling to their right or left, and charging with impetuosity the moment they take the enemy in flank. The first firing of the regular troops march out briskly and attack the enemy in front. The platoons detached in that manner from the two short faces proceed only 100 yards from their front where they halt to cover the square, while the rest of the troops who have attacked, pursue the enemy till they are totally dispersed, not giving them time to recover themselves.

APPENDIX V

REGIMENTAL MUSIC

DURING the period at which the Royal American Regiment was raised the instruments in use by military bands appear to have comprised 2 Clarinets, 2 Oboes, 2 Bassoons, and 2 Horns in addition to Drums and other Instruments of Percussion. Later on Flutes and Serpents were added to the list; and at a still later period the Trombone, Key-bugle and Opheclide made their appearance. The original tune for the 60th was the 'Grenadiers' March,'[1] the date of whose composition is lost in obscurity, but which appears to have formed the foundation of the modern march 'The British Grenadiers.' The proof that this was the march of our Regiment is given in a letter dated June 17, 1763, to Colonel Bouquet, written by Captain Lewis Ourry, at that time besieged by the Indians at Fort Bedford. He says: 'I long to see my native scouts come in with intelligence but I long more to hear the "Grenadiers' March."' The 'Grenadiers' March' or an adaptation of it seems to have been the regimental tune for quite half a century. A correspondent writing from Ramsgate on August 28, 1807, says: 'The 5th Battalion of the 60th advanced to the pier with the music playing the "British Grenadiers."' He adds that 'the pier was crowded with fashionable company which the music and choruses of the troops delighted.'

In the British Museum may be found 'A set of Marches for 2 Clarinets, Hautboys or German Flutes, 2 Horns and a Bassoon,' inscribed 'To the Right Honourable Lady Amherst, by I. R. Printed and sold by R. Bremner in the Strand, price 5*s*.' It is conjectured that I. R. was General John Reed.

Of these two marches one was composed 'for the 1st Battalion, 60th Regiment, General Haldimand's'; and the other 'for the 2nd Battalion, Colonel Christie's.' The names of General Haldimand and Colonel Christie show that they must have been composed

[1] This seems to have been the march of all infantry regiments at this period.

not earlier than 1778 and not later than 1781. They are printed at the end.

In the year 1796 the 'Duke of York's March' was composed, and later on—possibly in the following year when the Duke became our Colonel-in-Chief—was introduced as the Slow March of the Regiment. The Slow March was however, abolished more than forty years ago, and has latterly been played only at Inspections.

Weber's opera *Der Freischütz* came out about the year 1820, and 'The Huntsmen's Chorus' from that Opera was adopted as the Regimental March both by the Rifle Brigade and ourselves. Thenceforward until 1905, 'The Huntsmen's Chorus' seems to have been continuously played by the 2nd Battalion.[1]

Major G. H. Courtenay, who did not leave the 1st Battalion until 1842, told the writer that he was convinced that this and no other tune was in use during the whole time that he was in the Regiment; but a very few years later a march called 'The Poacher' was introduced. It appears to have lasted, however, a very short time. Colonel George Bliss MacQueen states that when he joined the 1st Battalion in 1849 the march in vogue was Lutzow's 'Wilder Yagd,' and had been so at all events since 1845. Nevertheless for some years this march although used concurrently with the 'Huntsmen's Chorus' does not appear to have entirely superseded it.

But the rhythm of Lutzow's 'Wild Chase'—also set to music by Weber—was however found unsuitable for our marching: naturally enough, for it had been composed as a cavalry march. It is said that the march was handed to an enterprising Bandmaster, who undertook to correct the difficulty, which he did by eliminating all but a few bars of Lutzow, and substituting Von Gehriech's 'Yägersleben,' while still retaining the name of 'The Wild Chase.'

About the year 1855 'The Huntsmen's Chorus' disappeared entirely from the 1st Battalion; and the 3rd Battalion, which was raised in that year at first took the modernised Lutzow's 'Wild Chase' as its march. The 4th Battalion, raised in 1858, adhered however, to 'The Huntsmen's Chorus,' which was adopted by the 3rd Battalion in 1861.

The 1st Battalion continued to use the modernised 'Wild Chase,' but the residual elements of the real tune, being still found inconvenient, were eliminated sometime about the year 1870; and

[1] Major Courtenay said that the translation of the Chorus began with the words 'Oh! sight entrancing'—a natural exclamation of the spectator!

thenceforward although still retaining the old name, the march was in fact entirely the 'Yägersleben,' the rhythm of which was extremely good for the step of Riflemen, better indeed than that of 'The Huntsmen's Chorus.' In 1882 'The Huntsmen's Chorus' was re-introduced in the 1st Battalion, and became once more the march of the whole Regiment until 1905, when—as would seem through a misapprehension of historical fact—it was again superseded—this time throughout the Regiment—by the 'Yägersleben' under the title of 'Lutzow's Wild Chase,' some bars of the original Lutzow air being however re-introduced and played by the Buglers.

While speaking of regimental music a word should perhaps be said of our Buglers. The Buglehorn-man, as he was then termed, was introduced into the Regiment with the 5th Rifle Battalion in 1798. The French General Foy tells us that 'the echoing sound of the Riflemen's horns answered the double purpose of directing their movements and of signalling such movements of the enemy as would otherwise have escaped the notice of the General-in-Command.'

Of late years the proficiency of our Buglers has further increased: the performance of those of the 4th Battalion at the Royal Military Tournament in 1909 was the subject of general comment and admiration; and those who were present in Winchester Cathedral at the dedication of the Memorial Window and the unveiling of the Buller monument, in the two following years, cannot fail to recall the beautiful rendering of the 'Last Post' by the buglers of the 1st Battalion, and the echo of the touching notes floating through the aisles.

In regard to vocal music, no tradition remains of the men's choruses referred to above. It is however known that in 1840 the song 'Over the Garden Wall' was a favourite among the Riflemen, and it has been stated that they used to sing it as far back as the Peninsular War.

A SET OF MARCHES

FOR TWO

CLARINETS, HAUTBOYS, OR GERMAN FLUTES, TWO HORNS & A BASSOON.

INSCRIBED TO

THE RT. HONBLE. LADY AMHERST.

BY I. R., ESQRE.

N.B.—For the ease of Harpsichord players, the first or principal part is placed next that of ye Bass.

LONDON. Printed & sold by R. BREMNER in the Strand.

Pr. 5*s.*

March for the First Battalion, 60th Regiment.
General Haldimand's.

March for the Second Battalion, 60th Regiment.
Colonel Christie's.

tr
tr
tr
tr

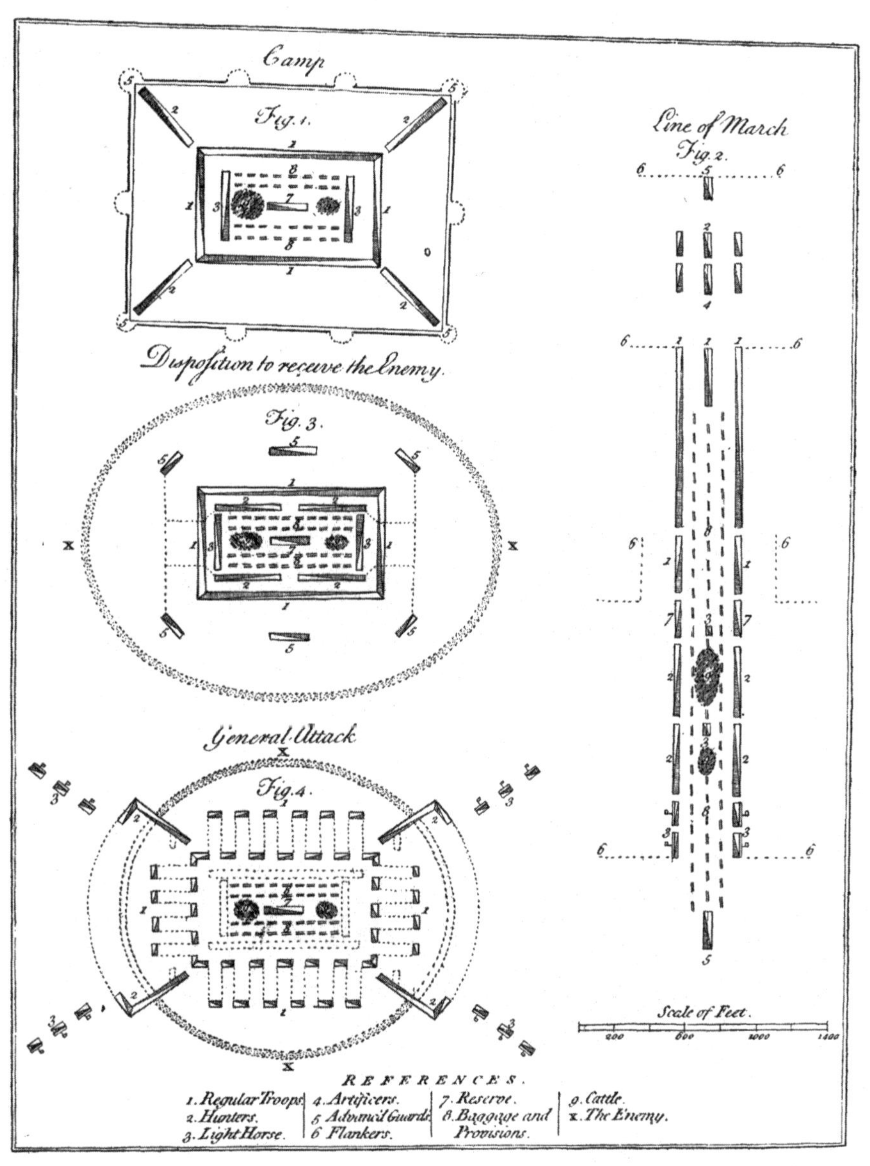

TACTICAL FORMATIONS FOR INDIAN WARFARE

APPENDIX VI

LISTS OF THE OFFICERS AT VARIOUS DATES

1756

Colonel-in-Chief

John, Earl of Loudoun . . . Dec. 25, 1755

Colonels-Commandant

1 John Stanwix .	. Jan. 1, 1756	3 Charles Jefferys	. Jan. 3, 1756
2 Joseph Dusseaux	. Jan. 2, 1756	4 James Prevost .	. Jan. 4, 1756

Lieutenant-Colonels

1 Henry Bouquet	. Jan. 3, 1756	3 Russell Chapman	. Jan. 5, 1756
2 Frederick Haldimand	Jan. 4, 1756	4 Sir John St. Clair	. Jan. 6, 1756

Majors

1 John Young .	. Dec. 25, 1755	4 John Rutherford	. Jan. 6, 1756
2 James Robertson	. Dec. 20, 1755	3 Augustin Prevost	. Jan. 9, 1756

Captains

1 John Tullikens	. Dec. 25, 1755	3 Harry Charteris	. Jan. 8, 1756
2 Thos. Oswald .	. Dec. 26, 1755	4 Lewis Stiener .	. Jan. 10, 1756
3 Rodolf Faesch .	. Dec. 27, 1755	1 Francis Lander	. Jan. 11, 1756
4 Fredk. Porter .	. Dec. 28, 1755	2 —— Rollaz .	. Jan. 12, 1756
1 Baron D. H. Munster	Dec. 29, 1755	3 John Innis .	. Jan. 13, 1756
2 Walter Rutherford	. Dec. 30, 1755	4 —— Schrader .	. Jan. 14, 1756
3 —— Wittsteen.	. Dec. 31, 1755	1 Gavin Cochran	. Jan. 15, 1756
4 Thos. Graeme .	. Jan. 1, 1756	2 Joseph Prince .	. Jan. 16, 1756
1 Ralph Harding	. Jan. 2, 1756	3 Marcus Prevost	. Jan. 17, 1756
2 —— Chambier .	. Jan. 3, 1756	4 Thos. Stanwix .	. Jan. 18, 1756
3 Jeremiah Stanton	. Jan. 4, 1756	1 Alex. Harbord .	. Jan. 19, 1756
4 —— Knielling .	. Jan. 5, 1756	2 Abraham Bosomworth	Jan. 20, 1756
1 Richd. Mather .	. Jan. 6, 1756	3 John Faesch .	. Jan. 21, 1756
2 Gust. Wetterstroom	. Jan. 7, 1756		

Captain-Lieutenants

1	—— Konn	Dec. 25, 1755	3	Stephen Gualley	Dec. 27, 1756
2	John Dalrymple	Dec. 26, 1756	4	Edwin Comberbach	Dec. 28, 1756

Lieutenants

1	Gilbert M'Adam	Dec. 26, 1755	3	John Sealy	Jan. 28, 1756
*2	Joseph Schlosser	Dec. 27, 1755	*4	Thos. Grandidier	Jan. 29, 1756
3	Charles Crookshanks	Dec. 28, 1755	1	James Campbell	Jan. 30, 1756
*4	Saml. Jan Hollandt	Dec. 29, 1755	2	Simon Frazer	Jan. 31, 1756
1	George Brereton	Dec. 30, 1755	3	George Fullerton	Feb. 1, 1756
2	Charles Forbes	Dec. 31, 1755	4	Wm. Stuart	Feb. 2, 1756
3	Francis Pringle	Dec. 31, 1755	1	Alex. Campbell	Feb. 3, 1756
4	Robt. Brigstock	Jan. 1, 1756	2	Joseph Ray	Feb. 4, 1756
*1	Peter von Ingen	Jan. 2, 1756	3	George Turnbull	Feb. 5, 1756
2	Alex. M'Bean	Jan. 3, 1756	4	Wm. Abercrombie	Feb. 6, 1756
3	Donald Campbell	Jan. 4, 1756	1	David Ouchterlony	Feb. 7, 1756
4	Newsham Piers	Jan. 5, 1756	2	Wm. Hazlewood	Feb. 8, 1756
*1	—— Habberthorn	Jan. 6, 1756	3	John Brown	Feb. 9, 1756
2	John Longsden	Jan. 7, 1756	4	Daniel M'Alpin	Feb. 10, 1756
3	Allan M'Lean	Jan. 8, 1756	1	Donald Forbes	Feb. 11, 1756
4	George Warburton	Jan. 9, 1756	2	Henry Gordon	Feb. 12, 1756
*1	—— Barnstedt	Jan. 10, 1756	3	—— Mackay	Feb. 13, 1756
2	Cunningham Cooper	Jan. 11, 1756	4	Thos. Basset	Feb. 14, 1756
3	Bazil Dunbar	Jan. 12, 1756	*1	George Faesch	Feb. 15, 1756
4	Robt. Drew	Jan. 13, 1756	2	George Etherington	Feb. 16, 1756
1	Louis Ourry	Jan. 14, 1756	*3	Emanuel Hesse	Feb. 17, 1756
2	James Dalzell	Jan. 15, 1756	*4	Rodolphus Bentinck	Feb. 18, 1756
3	Richd. Wynne	Jan. 16, 1756	*1	John Muller	Feb. 19, 1756
*4	James Allaz	Jan. 17, 1756	*2	Bernard Ratzer	Feb. 20, 1756
1	Ebenezer Warren	Jan. 18, 1756	*3	Dietrich Brehm	Feb. 21, 1756
2	William Baillie	Jan. 19, 1756	*4	Fred. von Weissenfels	Feb. 22, 1756
3	John Swift	Jan. 20, 1756	*1	Jos. Fredk. Wallet des Barres	Feb. 23, 1756
*4	Christr. Spiesmacher	Jan. 21, 1756	*2	Conrad Gugy	Feb. 24, 1756
1	William Cook	Jan. 22, 1756	*3	—— Kleinbeil [1]	Feb. 25, 1756
*2	Elias Meyer	Jan. 23, 1756	*4	—— Zimmerman [1]	Feb. 26, 1756
3	Henry Simcocks	Jan. 24, 1756	1	Lewis Val. Fuser	Feb. 27, 1756
4	Simon Ecuyier	Jan. 25, 1756	2	A. T. F. Winter	Feb. 28, 1756
1	Charles Willington	Jan. 26, 1756	*3	J. Von Ingen	Feb. 29, 1756
2	Charles Gallot	Jan. 27, 1756			

Ensigns

1	Bereton Poynton	Dec. 25, 1755	2	Ralph Phillips	Dec. 29, 1755
2	James Allen	Dec. 25, 1755	3	Saml. M'Kay	Dec. 30, 1756
3	Thos. Barnsley	Dec. 26, 1755	4	Francis M'Kay	Dec. 31, 1756
4	George M'Intosh	Dec. 27, 1755	1	George Archbold	Jan. 1, 1756
1	Thos. Campbell	Dec. 28, 1755	2	James Monro	Jan. 2, 1756

[1] Officers named Panier and Desmoilles appear to have been substituted for Kleinbeil and Zimmerman.

Those marked * seem to be the Engineers.

Ensigns (continued)

3 Wm. Ridge	Jan. 3, 1756	4 Abraham Hart	Jan. 16, 1756
4 Wm. Hay	Jan. 4, 1756	1 Robt. Campbell	Jan. 17, 1756
1 Alex. Shaw	Jan. 5, 1756	2 James Crofton	Jan. 18, 1756
2 Thos. Meredith	Jan. 6, 1756	3 Townshend Guy	Jan. 19, 1756
3 Stair Campbell Carre	Jan. 7, 1756	4 James Jeffries	Jan. 20, 1756
4 Walter Kennedy	Jan. 8, 1756	1 Francis Hutchinson	Jan. 21, 1756
1 Michael Davis	Jan. 9, 1756	2 James Herring	Jan. 22, 1756
2 Wm. Potts	Jan. 10, 1756	3 Edward Jenkins	Jan. 23, 1756
3 Andrew Watson	Jan. 11, 1756	4 James Ralfe	Jan. 24, 1756
4 Wm. Jones	Jan. 12, 1756	1 John Utterville	Jan. 25, 1756
1 John Bell	Jan. 13, 1756	2 Edward O'Brien	Jan. 26, 1756
2 Nicholas Sunderland	Jan. 14, 1756	3 Allen Grant	——
3 Wm. Ryder	Jan. 15, 1756	4 Alex. Grant	——

Chaplains

Thos. Gawton	Dec. 25, 1755	Wm. Nicholson Jackson	Feb. 4, 1756

Surgeons

George Tuting	Feb. 1, 1756	George Weigman	Feb. 3, 1756
John M'Kenzie	Feb. 2, 1756	—— Stevenson	Feb. 4, 1756

Agent

Mr. Caltraft, Channel Row, Westminster.

1767

Colonel-in-Chief

Sir Jeffery Amherst . . . . Sept. 30, 1758
Rank in army, Lieut.-General . Jan. 19, 1761

Colonels-Commandant

Hon. James Murray	Oct. 24, 1759	*James Prevost	Oct. 28, 1761
Rank in army, Major-General	July 10, 1762	Rank in army, Major-General	June 3, 1762

Lieutenant-Colonels

*Fred. Haldimand	Jan. 4, 1756	*Augustin Prevost	Dec. 13, 1765
Rank in army, Colonel	Feb. 19, 1762	Rank in army, Colonel	Mar. 20, 1761

Majors

Hon. Luc. Ferd. Cary	April 4, 1765	Robert Bayard	Oct. 4, 1765
Rank in army	Feb. 8, 1762		

* Joined Regiment 1756.

Captains

Name	Date
*Gavin Cochran	Jan. 15, 1756
John Bradstreet	Mar. 8, 1757
Rank in army, Col.	Feb. 17, 1762
*John Jos. Schlosser	July 20, 1758
*George Etherington	April 17, 1759
*Thomas Barnsley	May 30, 1759
*Sam. Jan Hollandt	Aug. 24, 1759
*Lewis Val. Fuser	Dec. 30, 1763
Rank in army	Sept. 27, 1762
*John Brown	June 14, 1764
Rank in army	Sept. 15, 1760
Stephen Kemble	Jan. 24, 1765
John Wharton	June 19, 1765
Rank in army	Sept. 17, 1760
Courtland Schuyler	Nov. 8, 1765
Rank in army	Jan. 21, 1760
*George Turnbull	Nov. 15, 1765
John Carden	Dec. 25, 1765
Rank in army	Feb. 25, 1760
Beamsley Glazier	Dec. 25, 1765
Rank in army	June 16, 1760

Captain-Lieutenants

Name	Date
Ralph Phillips	Dec. 12, 1760
*Fred. Christopher Spiesmacher	July 13, 1761

Lieutenants

Name	Date
*Thos. Grandidier	Jan. 29, 1756
*David M'Alpin	Feb. 10, 1756
*Bernard Ratzer	Feb. 20, 1756
*Dietrich Brehm	Feb. 21, 1756
*Jos. Fred. Wallet des Barres	Feb. 22, 1756
Ja. Von Ingen	Feb. 29, 1756
*Brereton Poynton	Nov. 30, 1756
*George M'Intosh	Dec. 3, 1756
*George Archbold	Dec. 8, 1756
John Polson	May 5, 1757
*Allan Grant	Oct. 7, 1763
Rank in army	July 28, 1758
*Stair Campbell Carre	May 8, 1764
Rank in army	May 7, 1757
*Francis Hutchinson	May 9, 1764
Rank in army	Aug. 23, 1758
Augustin Prevost	May 10, 1764
Rank in army	May 6, 1761
John Christie	Nov. 15, 1765
John Nordberg	Mar. 29, 1766
Rank in army	July 28, 1758
Lewis de Mestrall	Sept. 13, 1766
Rank in army	Mar. 31, 1760
Thomas Etherington	Sept. 13, 1766
Rank in army	May 27, 1760

Ensigns

Name	Date
Francis Schlosser	Aug. 29, 1759
Donald M'Donald	Feb. 14, 1760
John Amiel	June 26, 1760
Angus M'Donald	July 8, 1760
Christopher Pauli	Feb. 8, 1761
James Brigstock	April 24, 1761
George Price	May 18, 1761
Frederick Winter	July 7, 1761
Thomas Hutchins	Mar. 2, 1762
Matthew Keough	Oct. 31, 1763
Robert Johnson	April 25, 1765
Rank in army	April 20, 1761
Thomas Delamaine	Nov. 15, 1765
James Christie	Dec. 25, 1765
Rank in army	Oct. 19, 1762
Jn. Geo. Goldfrap	Mar. 21, 1766
David Alex. Grant	Sept. 3, 1766
Charles Duncan	Sept. 3, 1766
John K. Muller	Oct. 11, 1766

Chaplains

Name	Date
W. Nich. Jackson [1]	Feb. 4 1756
William Winder	April 4, 1765

* Joined Regiment 1756.

[1] In the *Army List* of 1756 this gentleman's name is given as 'Johnson.'

Adjutants

Patrick M'Alpin	. April 26, 1760	William Potts .	. 1762

Quartermasters

John Peter Rochat	. May 1, 1760	Francis Hutchinson .	Oct. 4, 1760

Surgeons

*James Stevenson	. Feb. 4, 1756	John Graham .	. March 12, 1766

Agent

Mr. Ross, Conduit Street.

1777

1st BATTALION

Colonel-in-Chief

Jeffery, Lord Amherst, K.B. .	. Nov. 7, 1768
Rank in army, Lieut.-General	. Jan. 19, 1761

Colonel-Commandant

*Frederick Haldimand . .	. Jan. 11, 1776
Rank in army, Major-General	. May 25, 1772

Lieutenant-Colonel

Gabriel Christie	. Sept. 18, 1775
Rank in army	. Jan. 27, 1762

Major

Stephen Kemble	. Sept. 20, 1775
Rank in army	. Aug. 7, 1772

Captains

*Fred. Chr. Spiesmacher	Oct. 4, 1770	Ralph Phillips .	. May 25, 1771
*George Archbold	. Jan. 23, 1771	*Bernard Ratzer	. Nov. 13, 1771
*Brereton Poynton	. April 13, 1771	John Polson .	. June 16, 1775
		*Jos. Wallet Des Barres	Sept. 23, 1775

Captain-Lieutenant and Captain

Chas. Dixon	. Nov. 12, 1776

Lieutenants

James Bain .	. May 6, 1772	Francis Duffield	. Nov. 13, 1772
Rank in army	. Dec. 11, 1762	Jeffrey Amherst	. April 12, 1773
David Alex. Grant	. May 11, 1772	Samuel Rutherford	. Feb. 16, 1774

* Joined Regiment, 1756.

Lieutenants (continued)

Spencer Briscoe	. Sept. 23, 1775	John Peter Rochat	. Sept. 30, 1775
George Hallam	. Sept. 28, 1775	Robert Palmer	. Nov. 12, 1776
George Browne	. Sept. 29, 1775	Charles Butter	. Nov. 13, 1776

Ensigns

James Fahy .	. May 6, 1773	William Mair .	. Sept. 29, 1775
Hans Carden .	.	Thomas Sentence	. Sept. 30, 1775
Rank in army	. June 16, 1773	John Henry Wolff	. Mar. 1, 1776
Thos. Milward Smith	Sept. 20, 1775	Simon Ecuyier .	. Mar. 13, 1776

Chaplain

James Mansety . . . . . May 18, 1776

Adjutant

Jeffery Amherst . . . . . Nov. 21, 1776

Quartermaster

John Peter Rochat . . . . May 1, 1760

Surgeon

Peter Walsh . . . . . April 29, 1767

2ND BATTALION

Colonel-in-Chief

Jeffery, Lord Amherst K.B. . . Nov. 7, 1768
Rank in army, Lieut-General . Jan. 19, 1761

Colonel-Commandant

*James Robertson . . . . Jan. 11, 1776
Rank in army . . . . May 25, 1772

Lieutenant-Colonel

*George Etherington . . . . Sept. 19, 1775

Major

*James Mark Prevost . . . Sept. 21, 1775
Rank in army (North America) . July 23, 1772

Captains

Francis Hutchinson	. Mar. 27, 769	*Dietrich Brehm	. Nov. 16, 1774
*Daniel M'Alpin	. Aug. 7, 1771	William Kelly .	. April 28, 1775
*Thomas Grandidier	. May 25, 1772	*James von Ingen	. Sept. 24, 1775
Thomas Etherington	Aug. 14, 1772		

* Joined Regiment, 1756.

Captain-Lieutenant and Captain

*John K. Muller . . Nov. 13, 1776

Lieutenants

Benjamin Wickham	June 3, 1772	Thomas Flucher	Oct. 4, 1774
Rank in army	Sept. 26, 1762	James Wakeley	April 18, 1776
Peter Graham	Jan. 6, 1773	Rank in army	Oct. 28, 1760
Marcus Pietet	June 16, 1773	Charles Southby	Aug. 14, 1776
John Chas. Schlosser	Nov. 16, 1774	Richard Hansard	Nov. 14, 1776
Poyntz Ricketts	Oct. 1, 1774	Thomas Walker	Nov. 15, 1776
John Bayard	Oct. 2, 1774		

Ensigns

David Gordon	Oct. 1, 1775	J. O. Degerman	May 17, 1776
John Charlton	Feb. 5, 1776	John Sharpe	Aug. 14, 1776
Richard Bennet	Feb. 6, 1776	Charles Crochley	Nov. 14, 1776
John Gottsched	Feb. 29, 1776	Philip Priddie	Nov. 15, 1776

Chaplain

William Winder . . . . April 4, 1765

Adjutant

John Charlton . . . . . Feb. 5, 1776

Quartermaster

John Fleming . . . . . Nov. 7, 1774

Surgeon

William Notter . . . . June 28, 1775

3RD BATTALION

Colonel-in-Chief

Jeffery, Lord Amherst, K.B. . . Aug. 25, 1775
Rank in army, Lieut.-General . . Jan. 19, 1761

Colonel-Commandant

John Dalling . . . . . Jan. 16, 1776
Rank in army . . . . May 25, 1772

Lieutenant-Colonel

William Stiell . . . . . Sept. 21, 1775

Major

*John Brown . . . . . Sept. 22, 1775

* Joined Regiment, 1756.

Captains

Jacob van Braam	Aug. 31, 1775	Donald M'Donald	Sept. 27, 1775
Rank in army	Sept. 19, 1761	Christopher Pauli	Sept. 29, 1775
Jacob Muller	Sept. 3, 1775	Augustin Prevost	Nov. 12, 1776
Rank in army	April 11, 1763	Rank in army	Sept. 23, 1775
George M'Intosh	Sept. 25, 1775	George Bruère	Nov. 15, 1776

Captain-Lieutenant

———

Lieutenants

Archibald Lamont	Sept. 2, 1775	James Robertson	Feb. 5, 1776
Rank in army	May 15, 1757	H. Christ. Fenner	Feb. 7, 1776
George Mackay	Sept. 3, 1775	Fred. de Gentzkow	Nov. 1, 1776
Rank in army	March 20, 1761	N. F. C. Lorkell	Nov. 21, 1776
George Sneyder	Sept. 5, 1775		
Rank in army	Sept. 27, 1762		

Ensigns

Lewis Mattay	Sept. 1, 1775	Ferdinand Brock	Feb. 8, 1776
James Gordon	Sept. 2, 1775	Isaac Hesselburg	Feb. 9, 1776
Robert Lawe	Sept. 3, 1775	James Finlay	Feb. 13, 1776
Imbert de Traytorrens	Feb. 7, 1776		

Chaplain

Michael Schlacter . . . . . Sept. 1, 1775

Adjutant

George Eberhard . . . . . Sept. 1, 1775

Quartermaster

Lewis Genevay . . . . . Sept. 1, 1775

Surgeon

John Sommers . . . . . Nov. 10, 1775

4TH BATTALION

Colonel-in-Chief

Jeffery, Lord Amherst, K.B. . . Aug. 25, 1775
Rank in army, Lieut.-General . Jan. 19, 1761

Colonel-Commandant

*Augustin Prevost . . . . . Sept. 18, 1775

* Joined Regiment 1756.

Lieutenant-Colonel

*Lewis Val. Fuser . . . . . Sept. 20, 1775

Major

Beamsley Glazier . . . . . Sept. 23, 1775

Captains

*James Allaz .	. Sept. 1, 1775	John Christie .	. Sept. 26, 1775
Rank in army	. Oct. 2, 1761	Patrick Murray	. Sept. 28, 1775
*Simeon Ecuyier	. Sept. 2, 1775	William Wulff .	. Sept. 30, 1775
Rank in army	. April 22, 1762	Thomas Hutchins	. Nov. 13, 1776
*George Fæsch .	. Sept. 22, 1775	Rank in army	. Sept. 24, 1775

Captain-Lieutenant and Captain

Alexander Shaw . . . . . Nov. 14, 1776

Lieutenants

James Edwards	. Sept. 4, 1774	William Lechinwitz	. Oct. 4, 1775
Rank in army	. June 23, 1762	Harry Burrard	. Feb. 6, 1776
David Monins .	. Sept. 6, 1775	Fred. de Montrond	. March 20, 1776
Rank in army	. Jan. 14, 1763	John James Graham	Nov. 21, 1776

Ensigns

Alexander M'Donald	Sept. 2, 1775	John Campbell	. Feb. 10, 1776
George M'Kenzie	. Sept. 3, 1775	James Irwin .	. Feb. 11, 1776
†William Probick	. Sept. 3, 1775	Robert Lethbridge	. Feb. 12, 1776
Marsh Vaniper.	. Oct. 4, 1775	C. L. Theod. Schoedde	May 18, 1776

Chaplain

George Bowe. . . . . . . Feb. 5, 1776

Adjutant

Alexander Shaw . . . . . Sept. 1, 1775

Quartermaster

John Clarke . . . . . . Feb. 5, 1776

Surgeon

James Henderson . . . . . Nov. 10, 1775

Agent

Messrs. Ross and Gray, Conduit Street.

---

* Joined Regiment 1756.
† Or Porbeck, or Van Borbeck.

1788

Colonel-in-Chief

Jeffery, Lord Amherst, K.B. . . Nov. 7, 1768
Rank in army, General . . . Mar. 19, 1778

Colonels-Commandant

Fred. Haldimand, K.B. Oct. 20, 1772
Rank in army, Lieut-General Aug. 29, 1777
Gabriel Christie . May 10, 1786
Rank in army, Major-General Oct. 19, 1781
William Rowley . Oct. 3, 1787
Rank in army, Major-General Oct. 19, 1781
Hon. William Gordon Oct. 3, 1787
Rank in army, Major-General Oct. 19, 1781

Lieutenant-Colonels

Stephen Kemble . May 14, 1778
Rank in army, Colonel Nov. 20, 1782
Archibald M'Arthur . Sept. 24, 1787
Rank in army . April 24, 1781
Peter Hunter . . Sept. 24, 1787
Rank in army, Lieut.-Colonel Nov. 23, 1782
Jas. Adolphus Harris Jan. 16, 1788

Majors

Patrick Murray . Oct. 6, 1784
Wm. Gooday Strutt . Sept. 24, 1787
Rank in army . Aug. 9, 1783
John Moore . . Jan. 16, 1788
Rank in army . Nov. 23, 1785
George Benson . Jan. 16, 1788

Captains

*J. F. Wallet des Barres Sept. 23, 1775
Rank in army, Major Mar. 19, 1783
Benjamin Wickman . Aug. 10, 1777
Geo. Ferdinand Loup Oct. 6, 1784
Richard Porter . Nov. 26, 1784
Mark Pietet . . Feb. 26, 1785
Rank in army . Dec. 26, 1778
Rich. Massey Hansard Mar. 23, 1785
Edward Davis . . Mar. 23, 1785
Thomas Walker . Nov. 4, 1785
Lacklan Maclean . Feb. 8, 1786
Rank in army . Oct. 17, 1782
Duncan Mackintosh . June 2, 1786
Rank in army . April 29, 1783
John Parr . . July 5, 1786
Frederick de Diemar . Sept. 24, 1787
Rank in army . June 20, 1778
George Schneider . Sept. 24, 1787
Rank in army . Dec. 26, 1778
Frederick de Montrond Sept. 24, 1787
Rank in army . May 24, 1779
Hon. Charles Curzon . Sept. 24, 1787
Rank in army . Oct. 11, 1781
James Ecuyier . Sept. 24, 1787
Rank in army . Oct. 20, 1781
Samuel de Vismes . Sept. 24, 1787
Rank in army . Dec. 15, 1781
William Lachenwitz . Sept. 24, 1787
James Wakeley . Sept. 25, 1787
William Freemantle . Sept. 26, 1787
Rank in army . May 24, 1783
Alexander Philip Forbes Sept. 26, 1787
James Lowe . . Sept. 26, 1787
Charles Ingram . Sept. 26, 1787
Henry Thurloe Shadwell Sept. 27, 1787
Rank in army . Dec. 2, 1779

* Joined Regiment 1756.

Captains (continued)

Name	Date
Andrew Philip Skene	Sept. 27, 1787
Rank in army	March 29, 1783
William Martin	Sept. 27, 1787
Robert Lethbridge	Sept. 27, 1787
Fred. de Chambault	Jan. 16, 1788
Rank in army	Sept. 27, 1787

Captain-Lieutenants and Captains

Name	Date
David Gordon	Sept. 30, 1787
Archibald McDonald	Sept 30, 1787
Joseph Breitenbach	Sept. 30, 1787
Frederick Gottsched	Sept. 30, 1787

Lieutenants

Name	Date
Ch. Lewis Theo. Schedde	Dec. 25, 1778
William Van Borbeck[1]	Dec. 26, 1778
Isaac Hesselburg	Dec. 26, 1778
Charles Crotchley	Dec. 27, 1778
William Floyer	Mar. 17, 1779
Gabriel Gordon	Nov. 26, 1784
John Campbell	Nov. 26, 1784
John Lennox	Nov. 26, 1784
James Bruere	April 6, 1785
Robert Bowker Jebb	May 18, 1785
Robert Mackworth	May 3, 1786
Rank in army	Sept. 9, 1783
Samuel Kempthorne	Jan. 10, 1787
Rank in army	March 20, 1783
John Kearsley	Sept. 24, 1787
Rank in army	Sept. 3, 1779
Henry Crusay	Sept. 24, 1787
Rank in army	Sept. 21, 1779
Enoch Plummer	Sept. 24, 1787
Rank in army	Sept. 22, 1779
Robert Hollandt	Sept. 24, 1787
Rank in army	June 29, 1780
Charles Brown	Sept. 24, 1787
Rank in army	Sept. 23, 1780
Lorentz Greenholme	Sept. 24, 1787
Rank in army	Nov. 20, 1780
Anthony Pasquada	Sept. 24, 1787
Rank in army	June 25, 1781
Hubert von Hamele	Sept. 24, 1787
Rank in army	Aug. 15, 1781
Henry Graydon	Sept. 24, 1787
Rank in army	Oct. 22, 1781
Charles de Diemar	Sept. 24, 1787
Rank in army	Dec. 15, 1782
John Watson	Sept. 24, 1787
Rank in army	March 16, 1783
Samuel Willis	Sept. 24, 1787
James Barrick	Sept. 24, 1787
John Young	Sept. 25, 1787
Rank in army	May 15, 1780
Edward Cartwright	Sept. 25, 1787
Thomas Clarke	Sept. 25, 1787
Robert Adolphus Farmer	Sept. 25, 1787
Joseph Pasquada	Sept. 25, 1787
John Lewis Prevost	Sept. 25, 1787
John Manners Kerr	Sept. 25, 1787
John Johnes	Sept. 25, 1787
Frederick de Vos	Sept. 26, 1787
Philip Masseria	Sept. 27, 1787
John Robert Nason	Oct. 24, 1787
Willam Claus	Oct. 31, 1787
John Robert Nolon	Oct. 24, 1787
James Dodds	Nov. 7, 1787
Rank in army	May 1, 1781
William Robins	Dec. 5, 1787
Francis Guarme	Jan. 16, 1788

Ensigns

Name	Date
Dorrell Cope	April 6, 1785
William Tireman	April 15, 1785
John Randall Forster	Nov. 2, 1785
Rank in army	July 11, 1781
Jacob Jordan	Jan. 11, 1786
Henry Le Moine	Feb. 7, 1787
Rank in army	Oct. 3, 1781
Hendes Molesworth	June 28, 1787
Augustus Gould	Sept. 24, 1787
Rank in army	May 21, 1779
Alexander Fraser	Sept. 24, 1787
Rank in army	Oct. 17, 1782
Robt. Campbell McPherson	Sept. 24, 1787

[1] Called 'Probick' in 1777 List.

Ensigns (continued)

Robt. Chance Nixon . Sept. 26, 1787
John Dupuy . . Sept. 27, 1787
Lewis Muller . . Sept. 29, 1787
Lewis Schneider . Sept. 30, 1787
George Fourneret . Oct. 1, 1787
*Chas. Frederick Piguet Oct. 2, 1787
James Peachy . . Oct. 3, 1787
Samuel Hollandt . Oct. 4, 1787
Cavendish Nugent . Oct. 6, 1787
Robert Gordon . Nov. 7, 1787
Rank in army, 2nd Lieut. Dec. 3, 1781
Hamilton Faucett . Dec. 6, 1787
Alexander John Goldie Jan. 8, 1788
Daniel Nixon . . Jan. 16, 1788
Thomas Martin . Jan. 22, 1788
Rank in army . Sept. 25, 1787
Robert Tyler . . Jan. 23, 1788
John J. Visscher . Jan. 24, 1788

Chaplains

William Winder . April 4, 1765
George Bowe . . Feb. 5, 1776
James Manesty . Sept. 24, 1787
Charles Morgan (or Mongan) . . Sept. 24, 1787

Adjutants

Richard Massey Hansard Mar. 9, 1780
George Westphall . Nov. 26, 1784
Jas. Hamilton Gordon Dec. 12, 1787

Quartermasters

John Fleming . . Nov. 9, 1774
Robert Burton . April 12, 1786
John Sommers . Sept. 24, 1787
John Clarke . . Sept. 24, 1787

Surgeons

James Henderson . Nov. 10, 1775
James Wright . . April 29, 1784
John Clarke . . Sept. 24, 1787
Isaac Titford . . Sept. 24, 1787

NOTE.—During this year joined the Regiment Samuel Gibbs, who as a General Officer commanded an independent British Brigade in Germany during 1813 and was killed at New Orleans in 1815.

* Or Picquet, or Picquett.

APPENDIX VII

EXTRACTS FROM MUSTER ROLLS, 1757-1801

1st BATTALION

No. of Coys.	Place	Date	Period	Present		Absent		Casuals		Total	
				Officers	Men	Officers	Men	Officers	Men	Officers	Men
6	New York . .	13 May, 1765 .	25 Oct., 1763—24 April, 1764	4	53	14	314	—	—	18	367*
6	New York . .	13 May, 1765 .	25 April, 1764—24 Oct., 1764	9	61	9	228	—	—	18	289
9	Quebec, 5 coys.	18 July, 1767 .	—	—	—	—	—	—	—	—	—
—	Quebec, 1 coy.	18 May, 1767. .	25 Oct., 1766—24 April, 1767	14	237	17	123	—	—	31	370†
—	New York, 3 coys.	12 Aug., 1772 .	—	—	—	—	—	—	—	—	—
9	New York . .	12 Aug., 1772 .	25 April, 1767—24 Oct., 1767	3	85	28	306	—	—	31	391
10	Jamaica, Spanish Town	5 May, 1780 .	5 May, 1780	10	221	25	294	—	—	35	515‡
8	Jamaica, Spanish Town	24 June, 1780 .	6 May, 1780—24 June, 1780	8	142	20	255	1	20	29	417
10	Jamaica, Spanish Town	25 Dec., 1780 .	25 June—24 Dec., 1780 . .	12	161	23	243	—	—	35	404
11	Spanish Town.	5 Aug. 1783 .	25 Dec., 1782—27 June, 1783	15	439	21	51	—	80	36	570
11	Spanish Town.	24 Aug., 1783 .	—	not given	—	not given	—	—	—	not given	461
8	Jamaica, Spanish Town, 7 coys.	7 Feb., 1784 .	25 Aug., 1783—24 Dec., 1783	8	327	21	85	—	15	29	427
—	Clarendon, 1 coy.	26 March, 1784 .	—	—	—	—	—	—	—	—	—

* ? 1 man in Capt. Feash's Company.

† Including 10 men 'entered within the muster.'

‡ Date and place of muster of Capt. John Polson's Company not given.

2ND BATTALION

No. of Coys.	Place	Date	Period.	Present		Absent		Casuals		Total	
				Officers	Men	Officers	Men	Officers	Men	Officers	Men
9	See note 1 . .	See note 1 . . .	25 Oct., 1764—24 April, 1765	12	319	20	86	—	—	32	405
9	As note 1 . .	As note 1, but (a) 1 Nov., 1765 (c) 24 Oct., 1765 (e) 29 Oct., 1765	25 April, 1765—24 Oct., 1765	13	329	19	49	—	—	32	378
9	See note 2 .	See note 2 . . .	25 Oct., 1765—24 April, 1766	11	321	21	54	—	—	32	375
9	See note 3 .	See note 3 . . .	25 April, 1766—24 Oct., 1766	17	252	15	96	—	—	32	348
9	See note 3 .	See note 3 . . .	25 Oct., 1766—24 April, 1767	17	313	15	117	—	—	32	430

Note 1—(*a*) Capts. F. Haldimand's, S. Bradstreet's and Major L. Cary's Coys. were mustered at Crown Point, 28 Sept., 1765; (*b*) Capts. J. Brown's, S. Holland's and R. Brigstock's were mustered at Fort William Augustus on July 25, 1765 (Capt. R. Brigstock's coy. June 25); (*c*) Capt. R. Bayard's coy. at Fort George, New York, 12 Aug., 1765; (*d*) Capt. T. Barnsley's coy. at Fort Oswegatchie, 24 July, 1765; (*e*) Maj.-Gen. J. Murray's coy. at Ticonderoga, 3 Oct. 1765.

Note 2—4 coys. mustered at La Prairie, 28 May, 1766; Capt. G. Turnbull's coy. at Detroit, 26 July, 1766; Capt. J. Bradstreet's coy. at Fort Ontario, 24 Aug., 1766; Maj. R. Bayard's coy. at New York, 24 Apr., 1766; Lt.-Col. F. Haldimand's and Maj.-Gen. Murray's coys. at Fort Erie, 14 July 1766.

Note 3—(*a*) Capt. S. Holland's and Capt. J. Wharton's coys. at Niagara, 2 Aug., 1767; (*b*) Gen. Murray's and Col. Haldimand's coys. at Michilimackinac, 30 June, 1767; (*c*) Capt. T. Barnsley's coy. at Fort Erie, 1 Aug., 1767; (*d*) Capt. J. Bradstreet's coy. at Fort Ontario, 21 May, 1767; (*e*) Capt. J. Brown's, Maj. P. Bayard's and Capt. G. Turnbull's coys. at Detroit, 26 July 1767.

2ND BATTALION (*continued*)

No. of Coys.	Place	Date	Period	Present		Absent		Casuals		Total	
				Officers	Men	Officers	Men	Officers	Men	Officers	Men
9	As note 3, but for Capt. Barnsley read Capt. Boyle Rocke	As note 3, but . (*a*) 18 May, 1768 (*b*) 1 Aug., 1768 (*c*) 27 May, 1768 (*d*) 6 May, 1768 (*e*) 18 June, 1768	25 April, 1767—24 Oct., 1767	13	318	19	66	—	—	32	384
9	See note 4 . .	See note 4 . . .	25 Oct., 1767—24 April, 1768	13	336	19	84	—	—	32	420
9	See note 5 . .	See note 5 . . .	25 April, 1768—24 Oct., 1768	16	332	18	63	—	—	34	395
9	See note 6 . .	See note 6 . . .	25 Oct., 1768—24 April, 1769	13	342	21	66	—	—	34	408
9	See note 7 . .	See note 7 . . .	25April, 1769—24 Oct., 1769	13	343	19	66	—	—	32	409

Note 3—See p. 358.

Note 4—(*a*) Capt. S. Holland's and Capt. J. Wharton's coys. at Niagara, 2 Aug., 1767 ; (*b*) Gen. Bigoe Armstrong's and Col. Haldimand's coys. at Michilimackinac, 1 Aug., 1768 ; (*c*) Capt. T. Barnsley's and Capt. J. Stevenson's coys. at Fort Erie, 27 May, 1768 ; (*d*) Capt. J. Bradstreet's coy. at Fort Ontario, 21 May, 1767 ; (*e*) Capt. J. Brown's, Major P. Bayard's and Capt. E. Turnbull's coys. at Detroit, 21 July, 1767.

Note 5—(*a*) Gen. Armstrong's and Col. Haldimand's coys. at Michilimackinac, 20 June, 1770, 24 Aug., 1770 ; (*b*) Maj. Bayard's, Capts. Turnbull's and Brown's at Detroit, 14 June, 1770 ; (*c*) Capt. Holland's, Capt. Stevenson's coys. at Niagara, 28 May, 1770 ; (*d*) Capt. Bradstreet's coy. at Fort Ontario, 15 May, 1770 ; (*c*) Capt. J. Wharton's coy. at Fort Erie, 3 June, 1770.

Note 6—(*a*) Gen. Armstrong's and Col. Haldimand's coys. at Michilimackinac, 30 June, 1770, and 24 Aug., 1770; (*b*) Capt. Stevenson's and Capt. Holland's at Niagara, 28 May, 1770 ; (*c*) Major Wharton's, Capts. Turnbull's and Brown's coys. at Detroit, 14 June, 1770 Major R. Bayard retired 17 March, 1769) ; *d*) Capt. J. Bradstreet's coy. at Fort Ontario, 15 May, 1770 ; (*e*) Capt. F. Hutcheson's coy. at Fort Erie, 3 June, 1770.

Note 7—As note 6, but Gen. Armstrong's coy. was mustered on 20 June, not the 30th. The Hon. Major Thos. Bruce was apptd. 9th June, 1769, vice Major John Wharton transferred to the 1st Bn., 8 June, 1769.

2nd BATTALION (*continued*)

No. of Coys.	Place	Date	Period.	Present		Absent		Casuals		Total	
				Officers	Men	Officers	Men	Officers	Men	Officers	Men
9	See note 8 . . .	See note 8 . .	25 Oct., 1769—24 Oct. 1769	15	356	19	67	—	—	34	423
9	See note 9 . . .	See note 9 . .	25 Dec., 1769—24 April, 1770	15	343	19	85	—	—	34	428
9	See note 10 . .	See note 10 .	25 April, 1770—24 Oct., 1770	10	276	22	141	—	—	32	417
9	See note 10 . .	See note 10 .	25 Oct., 1770—24 Dec., 1770	10	276	22	145	—	—	32	421
10	See note 10 . .	See note 10 .	25 Dec., 1770—25 Mar., 1771	12	273	22	147	—	—	34	420*
10	See note 10 . .	See note 10 .	26 Mar., 1771—24 April, 1771	14	279	20	138	—	—	34	417
10	New York . . .	24 Oct., 1772 .	25 April, 1771—24 Oct., 1771	10	189	25	189	—	—	35	378

Note 8—Gen. Armstrong's and Col. Haldimand's coys. at Michilimackinac, on June 20, 1770 and 24 Aug., 1770; Capts. Holland's and Bradstreet's coys. at Niagara, 28 May 1770; The Hon. Major Bruce's and Capts. Brown's and Turnbull's coys. at Detroit, 14 June, 1770; Capt. Stevenson's coy. at Fort Ontario, 15 May, 1770; Capt. F. Hutcheson's coy. at Fort Erie. 24 June, 1770.

Note 9—As note 8, but Capt. Stevenson's coy. at Niagara 26 June, 1770, Capt. Holland's and Bradstreet's coys. were also mustered on that date.

Note 10—Gen. Armstrong's and Col. Haldimand's coys. at Michilimackinac, 25 March, 1772; Maj.-Gen. Etherington's (Major Bruce preferred, 4 Oct., 1770), and Capts. Turnbull's and Brown's coys. at Detroit, 16 Aug., 1771; Capts. Hutcheson's, Stevenson's, Holland's and Bradstreet's coys. at Niagara, 22 July, 1771.

* Capt. B. Glazier's coy. mustered at Niagara, 23 Sept., 1771, is added.

2nd BATTALION (*continued*)

No. of Coys.	Place	Date	Period	Present		Absent		Casuals		Total	
				Officers	Men	Officers	Men	Officers	Men	Officers	Men
10	New York . . .	24 Oct., 1772 .	25 Oct., 1771—24 April, 1772	10	215	24	165	—	—	34	380
10	New York . . .	24 Oct., 1772 .	25 April, 1772—24 Aug., 1772	23	227	12	76	—	—	35	303§
3*	St. Augustine .	29 Jan., 1778 .	25 June, 1776—24 Dec., 1776	2	47	7	35	5	28	14	110
3*	St. Augustine .	29 Jan., 1778 .	25 Dec., 1776—24 June, 1777	2	56	8	36	4	15	14	107
3*	St. Augustine .	29 Jan., 1778 .	25 June, 1777—24 Dec., 1777	3	59	9	36	—	—	12	95
6	Antigua . . .	4 March, 1780.	25 Dec., 1777—23 Feb., 1780	14	169	7	122	—	10	21	301
10	Antigua . . .	25 Dec., 1780.	25 June, 1780—24 Dec., 1780	19	250	16	81	—	—	35	331
2†	Antigua . . .	25 Oct., 1782 .	25 June, 1781—24 June, 1782	6	105	0	0	—	7	6	112
10‡	Barbadoes . . .	1 Aug., 1783 .	Not given	14	308	1**	33**	19††	10††	34	411‡‡

* General J. Robertson's, Capt. F. Hutcheson's and Capt. J. Vaningen (died 9 Aug., 1777).
† Capt. Tisdell Cockell's and Capt. B. Wickham's.
‡ Each list is headed 'Effective Roll of Captain ——'s Coy. Barbados, 1 Aug. 1783.'
§ 64 men are mentioned as being received and entertained besides the 303.

** On duty.
†† On furlough.
‡‡ N.C.O.'s and men on command, 3; sick, 57.

3RD BATTALION

No. of Coys.	Place	Date	Period	Present		Absent		Casuals		Totals	
				Officers	Men	Officers	Men	Officers	Men	Officers	Men
9	Fort Edward	24 Oct., 1757 . .	25 April, 1757—24 Oct., 1757	30	413	11	325	—	—	41	738
10*	Dartmouth, Nova Scotia	6 May, 1758 . .	25 Oct., 1757—24 April, 1758	34	961	9	50	—	—	43	1011
10	Quebec . .	13 Oct., 1759 . .	25 April, 1759—24 Oct., 1759	25	403	21	489	—	—	46	892
10	Quebec . .	2 Oct., 1760 . .	25 Oct., 1759—24 April, 1760	27	454	18	366	—	—	45	820
10	Quebec . .	25 Nov., 1760 .	25 April, 1760—24 Oct., 1760	31	468	14	308	—	—	45	776
10	Quebec . .	14 May, 1761 . .	25 Oct., 1760—24 April, 1761	27	444	17	302	—	—	44	746
10	Camp on Staten Island	26 Oct., 1761 . .	25 April, 1761—24 Oct., 1761	37	531	8	158	—	—	45	689
10	New York .	23 Feb., 1764 .	25 Oct., 1761—24 April, 1762	4	29	39	589	—	—	43	618
10	New York .	23 Feb., 1764 .	25 April, 1762—24 Oct., 1762	6	30	28	388	—	—	34	418
10	New York .	23 Feb., 1764 .	25 Oct., 1762—24 April, 1763	8	34	26	338	—	—	34	372
6†	St. Augustine	12 Feb., 1778 . .	25 June, 1776—24 Dec., 1776	3	112	11	21	2	154	16	287
6‡	St. Augustine	30 Jan., 1778 . .	25 Dec., 1776—24 June, 1777	3	130	11	24	—	124	14	278
6	St. Augustine	30 Jan., 1778 . .	25 June, 1777—24 Dec., 1777	10	213	9	46	—	1	19	260
4	Newtown, Long Island	3 Sept., 1781 . .	25 Dec., 1777—24 June, 1778	2	57	5	38	8	131	15	226
4	Newtown, Long Island	3 Sept., 1781 . .	25 June, 1778—24 Dec., 1778	2	57	5	41	8	95	15	193
4	Newtown, Long Island	3 Sept., 1781 . .	25 Dec., 1778—24 June, 1779	3	61	5	43	8	86	16	190
4	Newtown, Long Island	3 Sept., 1781 . .	25 June, 1779—24 Dec., 1779	3	69	5	51	6	65	14	185
10	See note 1 .	See note 1 . . .	25 Dec., 1779—24 June, 1780	6	197	8	76	10	126	24	399

* Muster of Capt. J. Innes' coy. taken at Halifax, 6 May, 1758. † For one company the muster of casuals was taken at St. Augustine, 30 Jan., 1778.
‡ One company was vacant, but a muster of 40 casuals was taken.
Note 1—A muster of 6 companies was taken at Hampstead—5 on 15 March, 1783; 1 on 13 March, 1783. The muster of the other 4 was taken at Newtown, Long Island, on 3 Sept., 1781.

3RD BATTALION (*continued*)

No. of Coys.	Place	Date	Period	Present		Absent		Casuals		Totals	
				Officers	Men	Officers	Men	Officers	Men	Officers	Men
10	See note 1	See note 1	25 June, 1780—24 Dec., 1780	7	207	8	79	10	119	25	405
10	See note 2	See note 2	25 Dec., 1780—24 June, 1781	9	218	9	84	7	59	25	361
10	See note 2	As note 2, but for 3rd Sept., 1781, read 18th Ap., 1782	25 June, 1781—24 Dec., 1781	11	250	11	78	2	56	24	384
10	See note 3	See note 3	25 Dec., 1781—24 June, 1782	10	233	13	93	1	39	24	365
10	Hampstead	15 March, 1783	25 June, 1782—24 Dec., 1782	17	217	12	70	0	3	29	290
10	Hampstead, Long Island	Note 4. 3 July, 1783	25 Dec., 1782—24 June, 1783	14	226	16	49	—	—	30	275
10	Camp at Newtown Creek	8 Sept., 1783	25 June, 1783—8 Sept., 1783	13	196	20	50	—	—	33	246
Not given	Chatham Barracks	10 Sept., 1788	24 Sept., 1787—24 Dec., 1787	8	50	17	13	5*	16*	30	79
See note 5	Chatham Barracks	10 Sept., 1788	25 Dec., 1787—24 June, 1788	11	176	22	56	2†	23†	35	255
4	Dominica, Morne Avocat	4 Feb., 1789	25 June, 1788—24 Dec., 1788	6	144	0	1	—	—	6	145
See note 5	St. John's, Antigua	2 Feb., 1789	25 June, 1788—24 Dec., 1788	8	121	26	246	1‡	8‡	35	375§
2	Plymouth, Montserrat	18 Feb., 1789	25 June, 1788—24 Dec., 1788	3	67	—	—	—	—	3	67

Note 1—See page 362.'

Note 2—Muster of 6 coys. taken at Hampstead—5 on March 15, 1783; 1 on March 13, 1783. Muster of other 4 coys. taken at Newtown, Long Island, 3 Sept. 1781.

Note 3—Muster of 6 coys. taken at Hampstead—4 on March 15, 1783; 1 on March 15, 1781; 1 on March 13, 1783. Muster of other 4 coys. taken at Newtown, Long Island, on July 12, 1782.

Note 4—Date of muster of one coy. is given as 3 July, 1782.

Note 5.—Muster roll of whole Battn. so number of companies not given.

* Non-effective since 25 Dec. † Non-effective since 25 June 1788. ‡ Non-effective since 24 June. § Some coys. were at Dominica, some at Montserrat.

3RD BATTALION (*continued*).

No. of Coys.	Place	Date	Period	Present		Absent		Casuals		Total	
				Officers	Men	Officers	Men	Officers	Men	Officers	Men
See note 5	Monk's Hill, Antigua	22 Oct., 1789 . .	25 Dec., 1788—24 June, 1789	8	116	27	260	—	14*	35	390**
4	Prince Rupert's Head, Dominica	31 Dec., 1789. .	25 Dec., 1788—24 June, 1789	4	113	—	11	—	17*	4	141
3	Plymouth, Montserrat	10 Nov., 1789 .	25 Dec., 1788—24 June, 1789	3	58	—	—	—	1†	3	59
See note 5	Monk's Hill, Antigua	27 Feb., 1790. .	25 June, 1789—24 Dec., 1789	8	118	27	306	—	9‡	35	433**
2	Montserrat .	Not given . . .	25 June, 1789—24 Dec., 1789	—	—	—	—	—	—	3	67
4	Dominica. .	Not given . . .	25 June, 1789—24 Dec., 1789	—	—	—	—	—	—	5	141
2	Plymouth, Montserrat	10 Nov., 1789 .	25 Dec., 1788—24 June, 1789	7	58	—	11	—	—	} 8	199
4	Prince Rupert's Head, Dominica	31 Dec., 1789. .	25 Dec., 1788—24 June, 1789	—	113	—	—	1§	17§	}	
See note 5	Monk's Hill, Antigua	17 July, 1790. .	25 Dec., 1789—24 June, 1790	7	125	28	226	—	86‖	35	437
See note 5	Monk's Hill, Antigua	1 June, 1791 . .	25 June, 1790—24 Dec., 1790	9	76	26	234	—	55¶	35	365
See note 5	Monk's Hill, Antigua	2 July, 1791 . .	25 Dec., 1790—24 June, 1791	9	112	25	260	—	—	34	372**

Note 5—See page 363.
* Non-effective since 24 June. † Non-effective since June 25, 1789. ‡ Non-effective since 24 Dec. 1789. § Non-effective since 24 June 1789.
‖ Non-effective since 24 June 1790. ¶ Non-effective since 24 Dec. ** Some coys. were at Dominica, some at Montserrat.

3RD BATTALION (*continued*)

No, of Coys.	Place	Date	Period	Present		Absent		Casuals		Total	
				Officers	Men	Officers	Men	Officers	Men	Officers	Men
See note 5 (2 coys. at Montserrat) (2 at Tortola)	Monk's Hill, Antigua	1 March, 1792 .	25 June, 1791—24 Dec., 1791	13	140	21	185	—	7*	34	332
See note 5	St. John's Barracks, Antigua	25 Sept., 1792 .	25 Dec., 1791—24 June, 1792	9	118	26	169	—	16†	35	303§§
See note 5	Guernsey . .	3 Dec., 1794 . .	25 June, 1792—24 Dec., 1792	10	21	7	54	16‡	195‡	33	270
Not given	Guernsey . .	3 Dec., 1794 . .	25 Dec., 1792—24 June, 1793	10	22	10	54	14§	181§	34	257
See note 5	Guernsey . .	3 Dec., 1794 . .	25 June, 1793—24 Dec., 1793	10	22	12	53	9‖	153‖	31	228
See note 5	Guernsey . .	3 Dec., 1794 . .	25 Dec., 1793—24 June, 1794	12	26	13	79	5¶	19¶	30	124
See note 5	See note 6 .	See note 6 . . .	25 June, 1794—24 Dec., 1794	1	33	9	98	18**	532**	28	663
See note 5	Tobago . .	17 April, 1799 .	25 Dec., 1794—24 June, 1795	1	33	8	126	21††	465††	30	624
The	Chatham Roll	detachment of 60th	25 Dec., 1794—24 June, 1795	—	—	—	—	—	—	—	77‖‖
The	Chatham Roll	detachment of 60th	25 June, 1795—24 Dec., 1795	—	—	—	—	—	—	—	84‖‖
See note 5	Tobago . .	17 April, 1799 .	25 June, 1795—24 Dec., 1795	3	33	14	124	25‡‡	256‡‡	42	413

Note 5—See page 363.

Note 6—Date and place of muster not given. There is the following note: 'By the books of Col. Patterson of the Quarter Master General's office it appears that this Regiment embarked for the Leeward Islands in May 1795.' There is a muster roll of a detachment of the 60th Regt. called the Chatham Roll for the period given above—No. of Battn. not given—Officers 3, N.C.O.s and Men, 246. (Mainly a list of recruits.)

* Non-effective since 24 Dec.
† Non-effective since 24 June 1792.
‡ Non-effective since 24 Dec. 1792.
§ Non-effective since 24 June 1793.
‖ Non-effective since 24 Dec. 1793.
¶ Non-effective since 24 June 1794.
** Non-effective since 24 Dec. 1794.
†† Non-effective since 24 June 1795.
‡‡ Non-effective since 24 Dec. 1795.
§§ 1 coy. at Tortola.
‖‖ Mostly recruits.

3RD BATTALION (*continued*)

No. of Coys.	Place	Date	Period	Present		Absent		Casuals		Total	
				Officers	Men	Officers	Men	Officers	Men	Officers	Men
Not given	Tobago . .	17 April, 1799 .	25 Dec., 1795—24 June, 1796	4	33	20	122	28*	189*	52	344
See note 5	Tobago . .	17 April, 1799 .	25 June, 1796—24 Dec., 1796	7	113	22	275	24†	274†	53	662
See note 5	Tobago . .	17 April, 1799 .	25 Dec., 1796—24 June, 1797	8	119	22	331	24‡	245‡	54	695
See note 5	See note 7 .	Not given . . .	25 June, 1797—24 Dec., 1797	11	434	23	343	21§	150§	55	927
See note 5	Not given . See note 7 .	Not given . . . See note 7 . . .	25 Dec., 1797—24 June, 1798	See note 8	316	See note 8	571	See note 8	131‖	See note 8	1018
10	Tobago . .	24 July, 1798. .	25 June, 1798—24 July, 1798	—	—	—	—	—	—	66	1021¶
10	Tobago . .	24 Aug., 1798. .	25 July, 1798—24 Aug., 1798	—	—	—	—	—	—	65	1015
10	Tobago . .	24 Sept., 1798 .	25 Aug., 1798—24 Sept., 1798	—	—	—	—	—	—	64	997
10	Tobago . .	24 Oct., 1798 . .	25 Sept., 1798—24 Oct., 1798	—	—	—	—	—	—	63	969
10	Tobago . .	24 Nov., 1798 .	25 Oct., 1798—24 Nov., 1798	—	—	—	—	—	—	62	927
10	Not given. .	24 Dec., 1798. .	25 Nov., 1798—24 Dec., 1798	—	—	—	—	—	—	63	899
10	Tobago . .	24 March, 1799 .	24 Feb., 1799—24 Mar., 1799	—	—	—	—	—	—	57	862
10	Not given .	Not given . . .	25 Mar., 1799—24 April, 1799	—	—	—	—	—	—	57	845**
10	Not given .	Not given . . .	25 Aug., 1799—24 Sept., 1799	—	—	—	—	—	—	57	817**
10	Tobago . .	24 Dec., 1799. .	25 Nov., 1799—24 Dec., 1799	—	—	—	—	—	—	47	867††
10	Tobago . .	24 Jan., 1800 . .	25 Dec., 1799—24 Jan., 1800	—	—	—	—	—	—	47	868††
10	Tobago . .	24 Feb., 1800. .	25 Jan., 1800—24 Feb., 1800	—	—	—	—	—	—	49	868††
10‡‡	Tobago . .	24 March, 1800 .	25 Feb., 1800—24 Mar., 1800	—	—	—	—	—	—	48	861††

Note 5—See page 363.

Note 7—Date and place of muster not given. Grenada and Europe are the two places put against the names.

Note 8—In the proof table there are no officers given either as absent or present or non-effective. There seem to be 66 officers altogether, but whether absent, present, or non-effective is not stated.

* Non-effective since 24 June 1796.
† Non-effective since 24 Dec. 1796.
‡ Non-effective since June 24, 1797.
§ Non-effective since 24 Dec. 1797.
‖ Non-effective since 24 June 1798.
¶ 1 coy. at Grenada.

** Adjutant's rolls. Those for June, July, August, October and November have been omitted. The dates and places are not given in any of them. There are ten companies in each roll. General Rowley is Colonel and Lieut.-Colonel Ramsey, C.O.

†† Four coys. at Grenada.
‡‡ 12 recruits sent from Chatham Barracks not posted to companies.

3RD BATTALION (*continued*)

No. of Coys.	Place	Date	Period	Present		Absent		Casuals		Total	
				Officers	Men	Officers	Men	Officers	Men	Officers	Men
10	Tobago	24 April, 1800	25 Mar., 1800—24 April, 1800	—	—	—	—	—	—	49	871*
10	See note A	24 May, 1800	25 April, 1800—24 May, 1800	—	—	—	—	—	—	49	869†
10†	See note A	24 June, 1800	25 May, 1800—24 June, 1800	—	—	—			—	49	920
10†	See note A	24 July, 1800	25 June, 1800—24 July, 1800	—	—	—	—	—	—	49	748
10	See note B	24 Aug., 1800	25 July 1800—24 Aug., 1800	—	—	—	—	—	—	50	739
10	See note B	24 Sept., 1800	25 Aug., 1800—24 Sept., 1800	—	—	—	—	—	—	55	743
10	See note B	24 Oct., 1800	25 Sept., 1800—24 Oct., 1800	—	—	—	—	—	—	53	728
10	See note B	24 Nov., 1800	25 Oct., 1800—24 Nov., 1800	—	—	—	—	—	—	51	719
10	See note B	24 Dec., 1800	25 Nov., 1800—24 Dec., 1800	—	—	—	—	—	—	57	717
10	See note B	24 Jan., 1801	25 Dec., 1800—24 Jan., 1801	—	—	—	—	—	—	48	710
10	See note B	24 Feb., 1801	25 Jan., 1801—24 Feb., 1801	—	—	—	—	—	—	48	708
10	See note B	24 March, 1801	25 Feb., 1801—24 Mar., 1801	—	—	—	—	—	—	47	698
10	See note B	24 April, 1801	25 Mar., 1801—24 April, 1801	—	—	—	—	—	—	49	773
10	See note B	24 May, 1801	25 April, 1801—24 May, 1801	—	—	—	—	—	—	47	769
10	See note B	24 June, 1801	25 May, 1801—24 June, 1801	—	—	—	—	—	—	46	766
10	See note B	24 July, 1801	25 June, 1801—24 July, 1801	—	—	—	—	—	—	45	770
10	See note B	24 Aug., 1801	25 July, 1801—24 Aug., 1801	—	—	—	—	—	—	45	758
10	See note B	24 Sept., 1801	25 Aug., 1801—24 Sept., 1801	—	—	—	—	—	—	45	760
10	See note B	24 Oct., 1801	25 Sept., 1801—24 Oct., 1801	—	—	—	—	—	—	42	760
10	See note B	24 Nov., 1801	25 Sept., 1801—24 Nov., 1801	—	—	—	—	—	—	40	745
10	See note B	24 Dec., 1801	25 Nov., 1801—24 Dec., 1801	—	—	—	—	—	—	40	752

Note A.—Sworn at Tobago.
* Tobago and Grenada.

Note B.—Sworn at Grenada.
† 4 coys. at Grenada.

4th BATTALION

No. of Coys.	Place	Date	Period	Present		Absent		Casuals		Total	
				Officers	Men	Officers	Men	Officers	Men	Officers	Men
10	Camp near Albany	31 Oct., 1757 .	25 April, 1757—24 Oct., 1757	30	384	13	448	—	—	43	832
10	New York. . .	24 April 1758 .	25 Oct., 1757—24 April, 1758	32	694	11	316	—	—	43	1010
10	Lake George . .	24 Oct., 1758 .	25 April, 1758—24 Oct., 1758	18	486	26	418	—	—	44	904
10	Schenectady . .	14 May, 1758 .	25 Oct., 1758—24 April, 1759	31	586	15	250	—	—	46	836
10*	Fort Ontario . .	12 June, 1760 .	25 Oct., 1759—24 April, 1760	28	489	18	264	—	—	46	753
10	See note 1. . .	See note 1 . .	25 April, 1760—24 Oct., 1760	27	469	18	170	—	—	45	639
10	Montreal . . .	16 July, 1761 .	24 Oct., 1760—24 Dec., 1760	21	371	24	231	—	—	45	602
6	St. Augustine .	31 Jan., 1778 .	25 June, 1776—24 Dec., 1776	12	187	9	33	—	76	21	296
6	St. Augustine .	31 Jan., 1778 .	25 Dec., 1776—24 June, 1777	13	208	9	38	—	44	22	290
6	St. Augustine .	31 Jan., 1778 .	25 June, 1777—24 Dec., 1777	13	293	10	47	—	1	23	341
10†	Hampstead . .	15 March, 1783	25 Dec., 1780—24 June, 1781	13	174	11	40	11	102	35	316
10†	Hampstead . .	15 March, 1783	25 June, 1781—24 Dec., 1781	13	235	11	47	9	91	33	373
10‡	Hampstead . .	15 March, 1783	25 Dec., 1781—24 June, 1782	16	263	12	45	4	56	32	364
10§	Hampstead . .	15 March, 1783	25 June, 1782—24 Dec., 1782	18	271	14	48	—	12	32	331
10‖	Hampstead . .	3 July, 1783 .	25 Dec., 1782—24 June, 1783	13	192	20	31	—	—	33	223
10	Camp, New Town	8 Sept., 1783 .	25 June, 1783—8 Sept., 1783	12	163	21	39	—	—	33	202

Note 1—2 coys. at St. Vincent de Pauel, 12 Dec. 1760 and 2 Nov. 1760; 3 coys. at Jene Bone, 1 Dec. 1760; 1 coy. at the Isle Jesus, St. Vincent de St. Paul, 2 Dec. 1760; 1 coy. at the Jessu, St. Vincent de Poure, 2 Dec. 1760; 1 coy. at Tarbon, 5 Dec. 1760; 1 coy. at St. Rose, 2 Dec. 1760; 1 coy. at Latheney, 1 Dec. 1760.

* There is a note on the outside of two of the lists as follows: '10 co.'s at Theyda Falls and Fort Ontario the 24 April and the 11 and 12 June 1760.'

† Capt. Murray's coy. is called his Light Infantry Company; Capt. Burrard's coy. is called his Grenadier Company.

‡ (*a*) As for note †, but for Capt. Murray read Capt. Francis Erskine; (*b*) On the list of one Company the date 15 March 1782 is written *not* 15 March 1783.

§ Capt. Erskine's coy. is called his Light Infantry Company; Capt. Burrard's coy. is called his Grenadier Company.

‖ (*a*) Capt. Murray's coy. at Hampstead on 15 March 1783; Capt. de Montrond's coy. at Hampstead on 2 July 1783. (*b*) Capt. Burrard's coy. is called his Grenadier Company.

APPENDIX VIII

ACT OF PARLIAMENT 29 GEO. II., CAP. 5

AN ACT to enable His Majesty to grant Commissions to a certain number of Foreign Protestants who have served Abroad as Officers, or Engineers, to act and rank as Officers, or Engineers, in AMERICA only, under certain Restrictions and Qualifications.

WHEREAS . . . many foreigners being Protestants, have been induced by the encouragement offered to them by the said Act, to reside and settle in some of the said Colonies (and particularly in the Provinces of Maryland and Pensylvania) the natural born Subjects of which last-mentioned Province do in great Part consist of the People called Quakers, whose Backwardness in their own Defence exposes themselves, and that Part of America, to imminent Danger: And whereas, for the better Defence of the said Colonies it hath been proposed to raise a Regiment there, consisting of Four Battalions of one thousand men each, and to enlist as Soldiers in the said Regiment any of the said Foreign Inhabitants of the said Colonies, who, together with the Natives, shall voluntarily enter themselves in His Majesty's Service as Soldiers; which Foreigners cannot so well be raised or trained, without the Assistance of some Officers who are acquainted with their manners and language: And whereas it is expedient in the present juncture of affairs, to facilitate the speedy raising of such Regiment, and to enable a certain number of Foreign Protestants who have served abroad as officers, or Engineers, and thereby acquired experience and knowledge, and to serve and receive Pay as Officers in the said Regiment, or as Engineers in America; be it enacted . . . That all such Foreign Protestants who shall receive Commissions from His Majesty, His Heirs or Successors, to be Officers in the said Regiment, or to be Engineers in America . . . shall and may be enabled to serve and receive Pay as Officers in the said Regiment, or as Engineers in America.

PROVIDED, nevertheless, that the number of such Officers in the said Regiment shall not in the whole, at any Time, exceed Fifty; nor the number of Engineers in the whole, at any time, exceed Twenty.

PROVIDED ALSO, that the Colonel of the said Regiment shall be a natural-born subject, and not any Person naturalized or made Denizen.

PROVIDED ALSO, that no such Foreign Officer shall be enabled by this Act to serve as the Officer or Engineer in any Place, except America only;—

. . . .

APPENDIX IX

MORTS SUR LE CHAMP D'HONNEUR

1758	Hay, William, Lieutenant . .	Louisbourg
	Forbes, Charles, Capt.-Lieutenant	Ticonderoga
	Davis, Michael, Lieutenant .	,,
	Rutherford, John, Major . .	,,
	Hazlewood, William, Lieutenant .	,,
	Rohr, Charles, Ensign . .	Fort Duquesne
1759	Ochterlony, David, Captain .	Montmorenci Heights
	Kennedy, Walter, Lieutenant .	,, ,,
	de Witt, Peter, Lieutenant .	,, ,,
	Johnston, Samuel, Ensign . .	,, ,,
1762	Stewart, David, Ensign . .	Havana
	McDougal, John, Lieutenant .	,,
	An Officer, name unknown .	,,
1763	Campbell, Donald, Captain .	Détroit
	Holmes, Robert, Ensign . .	Miamis
	Jamet, John, Lieutenant . .	Michilimackinac
	Gordon, Francis, Lieutenant .	Venango
	Dow, John, Lieutenant . .	Bushy Run
1778	Macdonald, Donald, Captain .	Sunbury
1779	Brock, Ferdinand, Lieutenant .	Bâton Rouge
	Finlay, James, Lieutenant . .	Georgia
	Hosleton, Rowland, Lieutenant .	,,
	Wolff, William, Captain . .	,,
	Muller, Jacob, Brevet-Major .	,,
1780	Charlton, John, Lieutenant. .	Black River Expedition
	Jeserick, J. Phillip, Ensign .	,, ,, ,,
	Plees, Christian, Ensign . .	,, ,, ,,
	Scriven, J. Barclay, Ensign .	,, ,, ,,
	Prevost, J. Marc, Lieut.-Colonel .	Jamaica
	Fuser, Lieut.-Colonel, L.V. . .	St. Augustine

1793	Dunlop, George, Captain . .	Martinique
1794	Conway, P. Luay, Lieutenant .	Guadeloupe
	Cunningham, William, Lieutenant	,,
	Schneider, G. C., Lieutenant .	,,
	Cook, William, Lieutenant . .	,,
1795	Picquet, F. Frederick, Captain .	St. Vincent
1809	Dolling, James A., Captain. .	'The Saints'

This probably does not exhaust the list of those officers who died on service during the period given.

INDEX

S

www.ingramcontent.com/pod-product-compliance
Ingram Content Group UK Ltd.
Pitfield, Milton Keynes, MK11 3LW, UK
UKHW021840270726
14058UKWH00002B/249

9 781843 424451